Hydrology of Disasters

Hydrology of Disasters

Edited by

Ö. STAROSOLSZKY AND O.M. MELDER

Proceedings of the Technical Conference in Geneva
organized by the World Meteorological Organization
November 1988

British Library Cataloguing in Publication Data

Hydrology of disasters
  1. Disasters. Hydrological factors
  I. Starosolszky, Ö.   II. Melder, O.M.   III. World
  Meteorological Organization
  551.48

  ISBN 0–907383–50–5

Typeset in Plantin by Columns Design and Production Services Limited

Printed in Great Britain by Antony Rowe Ltd, Chippenham, Wiltshire

Contents

Secretary-General's Statement at the Opening of the Technical Conference on the Hydrology of Disasters
===

G.O.P. OBASI

World Meteorological Organization, Geneva, Switzerland

MR CHAIRMAN, my dear colleagues,

It is a great pleasure for me to speak to you this morning, for many reasons. In the first place, it offers me the opportunity of extending to all of you a very warm welcome to Geneva, to WMO and to this International Conference Centre, the place where the WMO Commission for Hydrology is presently holding its eighth session

Some of you have, of course, been here many times before. Indeed, I have already seen many familiar faces this morning – and I greet you as old friends. Others among you, we have the honour and pleasure of seeing for the first time. To you, I extend a welcome that is no less warm and no less sincere. I hope very much that if this is the first time we see you in a WMO meeting, it will not be the last.

Another reason why I am glad to say a few words to you before you begin your deliberations is to stress to you the importance which we in WMO attach to this conference. A glance at the list of participants shows that WMO is not alone in holding this view. The high level of representation demonstrates that both national and international bodies are aware of the importance and significance of this conference. With such a wealth of scientific talent and experience participating, the success of the conference seems assured.

Let me explain in a little more detail why WMO considers this to be an important conference. In the first place, any major international conference on hydrology must nowadays be classed as in important event, for it is no exaggeration to say that the whole economic and social structure of the modern world is dependent upon adequate water supplies and hence upon the skill and knowledge of hydrologists and water engineers in harnessing this resource. In this particular case, however, there are more specific reasons than this generalized statement. The destruction caused each year by tropical cyclones, typhoons and hurricanes is well known. In some developing countries they may nullify a significant part of the great effort being made towards economic and social progress. Then there is the devastation which results from a shortage of rainfall and the consequent conditions of drought which are very much in our minds at the moment. In other regions, such devastation may be due to the opposite effect – excessive rainfall and consequent flooding, sometimes resulting in dam and levée ruptures. Hydrological consequences of volcanic eruptions and the impact of accidental pollution are also events that require our utmost attention.

Losses from natural disasters appear to be increasing every year with the growth in population, and we must now also consider the complications introduced by climatic change. For these reasons and because of the difference in responses to natural disasters between rich and poor nations, increasing attention is being focused on the need for urgent action aimed at reducing the impact of disasters.

As many of you are aware, the General Assembly of the United Nations passed a resolution on 11 December 1987 designating the 1990s as a decade in which the international community, under the aegis of the United Nations, will give special attention to promoting international co-operation in the field of natural disaster reduction. The disasters stemming from perturbations of the fluid earth are among the most savage. WMO, with its responsibilities in meteorology and operational hydrology, is therefore well placed to play a leading role in the international decade for natural disaster reduction. The present conference is called upon to review the hydrological aspects of different types of disasters and to summarize the present knowledge in this field. The conference may wish to suggest action required on national or international levels to prevent harmful hydrological consequences of disasters, and measures required to protect population and property against such events. I can assure you that your views on this matter will be fully taken into account in the plan of action for the WMO contribution to the international decade.

Ladies and Gentlemen, that completes all that I wish to say on this occasion. I hope to have the pleasure of meeting you all personally under less formal circumstances, but let me conclude by saying that it is my earnest desire that this conference will be a success. Let me repeat also the welcome I extended at the outset to all participants. We shall do our best to make you feel at home and we hope that your visit to Geneva and to WMO will not only be useful for you, but also a pleasant personal experience.

Statement

HANS EINHAUS
UNDRO, Palais des Nations, 1211 Geneva, Switzerland

I AM INDEED most grateful to you Mr Chairman for this opportunity to pay tribute to the very close and productive ongoing co-operation between WMO and UNDRO. This applies not only to the full spectrum of meteorological and hydrological aspects of disaster mitigation in many parts of the uorld, but also includes the very significant role that WMO has played with us in the preparations for an ambitious new initiative, namely the International Decade for Natural Disaster Reduction (IDNDR) and it is to this subject that I intend in the main to confine my report.

Just to illustrate the mutually complementary nature of WMO/UNDRO co-operation, let me refer to one example. Since the initiation of the Typhoon Operational Experiment, TOPEX, in 1980, UNDRO has co-operated with WMO in the Warning Dissemination and Information Exchange component of this programme. In particular, UNDRO has helped with the development of a standard format for damage assessment, and new measures for public education and information. These efforts have led to significant improvements in public response, typhoon warning and damage assessment procedures.

The past year unfortunately had more than its usual share of major disasters, and some of the worst of these have been hydrological. While the most widespread and devastating disaster was the flooding in Bangladesh, hurricanes in the Caribbean and the Philippines showed the power of wind and water together.

All of these events emphasize the importance of disaster mitigation.

The term mitigation or reduction has been adopted, with reference to disasters, for two particular reasons: first because disaster prevention is too presumptuous – it is doubtful whether any practicable measures can achieve the complete avoidance of loss, which the word prevention implies. Second, the term mitigation includes not only the long-term, so-called preventive measures, but also the short-term protective actions which are generally described as preparedness measures. The International Decade is concerned with all of these and other disaster reduction measures.

What the large storms and floods of 1988 have shown yet again is that while natural disasters present a challenge for all nations, rich or poor, their effects are relatively more devastating for developing and the poorer countries. It is therefore all the more important that everything possible be done to reduce these effects, and it is to this end that I hope that WMO and UNDRO will concert their efforts over the coming years, in co-operation with several other agencies whose mandate includes disaster mitigation.

The main objectives therefore of the International Decade for Natural Disaster Reduction, as outlined in General Assembly resolution 42/169, are to develop measures for the assessment, prediction and mitigation of natural disasters through programmes of technical assistance and technology transfer, demonstration projects, and education and training, tailored to specific hazards and locations, and to evaluate the effectiveness of existing as well as new programmes in this field.

The goals specified in the resolution are:

(a) to improve the capacity of each country to mitigate the effects of natural disasters expeditiously and effectively, paying special attention to assisting developing countries in the establishment, when needed, of early warning systems;

(b) to devise appropriate guidelines and strategies for applying existing knowledge, taking into account the cultural and economic diversity among nations;

(c) to foster scientific and engineering endeavours aimed at closing critical gaps in knowledge in order to reduce loss of life and property;

(d) to disseminate existing and new information related to measures for the assessment, prediction, prevention and mitigation of natural disasters;

(e) to develop measures for the assessment, prediction, prevention and mitigation of natural disasters through programmes of technical assistance and technology transfer, demonstration projects and education and training, tailored to specific hazards and locations, and to evaluate the effectiveness of those programmes.

In proposing the International Decade, the General Assembly recognized that scientific and technical understanding of the causes and impact of natural disasters and of ways to reduce human and property losses has progressed to such an extent that a concerted effort to assemble, disseminate and apply this knowledge could have very positive effects, in particular for developing countries. An example of the usefulness of such action was the case of hurricane Gilbert in September 1988 in the Caribbean. As a result of considerable assistance received from WMO in monitoring, and from UNDRO with disaster management (which included 5 joint WMO/UNDRO training seminars), the impact of one of the greatest hurricanes ever recorded in the Caribbean was considerably mitigated. UNDRO is in the process of promoting similar disaster mitigation activities in other parts of the world, including in particular one for the South Pacific Island States, to be based in Fiji, where we plan also to work in close co-operation with the ongoing WMO programme.

For those of you who have not received up-to-date information about preparations for the International Decade, I would like to highlight the main activities to date and our future expectations. The Decade will not be formally launched until January 1990, and is therefore still in the preparatory stage. Planning activities which began early this year have been carried out at two levels, firstly by a Steering Committee composed of Executive Heads of major UN departments and designated senior officials of organization and agencies like UNDP, UNEP, UNCHS, WFP, FAO, UNESCO, WMO, World Bank, ITU, WHO, and IAEA and secondly by an international group of experts.

The Steering Committee is assisted in its work by a Working Group. WMO's representatives have been among the most active members of both these bodies.

What is the Role of Individual Organizations within the United Nations System? Each organization and specialized agency would, within its area of functional competence, prepare its own action plan for the Decade, under the control of its governing body. Once approved, these plans would be communicated to the Steering Committee so that the dangers of duplication of effort and gaps would be avoided.

It would be open to each organization to provide expertise to assist in the implementation of national and regional programmes.

Co-operation between specialized agencies in the framing of these action plans would be particularly desirable where there is considerable interaction between their areas of interest – for example, the effects of weather and climate upon agriculture, and of all three upon health and nutrition.

The main role of the United Nations i.e. UNDRO, as opposed to the specialized agencies, will be to provide a centre for the exchange of information. This role can also be described as one of facilitating the transfer of technology, because there is much knowledge already available whose existence is unknown to potential users. National activities can therefore be assisted by the publication of lists of bibliographies which have been compiled by various institutions and of a list of basic manuals, guides and reports which are available; by the holding of seminars and training courses; and by establishing and reinforcing easily accessible regional data centres on disasters.

Information exchange itself can be facilitated by, *inter alia*, the holding of seminars; the publication of regular progress bulletins (both technical/scientific and popular) and at an individual level by on-the-job training. The benefits of information exchange will be increased if it is appreciated that methods of operation suitable for technologically advanced countries are not always applicable in developing countries, and that therefore attempts must be made to adjust the original information or procedures to the conditions appropriate to developing countries.

The understanding, co-operation and participation of members of the public will be required if the Decade is to be a success, and the United Nations can, through its several agencies, etc., provide relevant and appealing information material for public and educational use.

The second body concerned with the planning phase of the Decade is the International Ad Hoc Group of Experts on the IDNDR. This is a panel of 25 eminent scientists and experts appointed by the Secretary-General, in consultation with the Executive Heads of Specialized Agencies like WMO, to assist him in developing an appropriate framework for the Decade. The Expert Group, under the chairmanship of Dr Frank Press, President of the US National Academy of Sciences, has already held two meetings, and two additional meetings have been tentatively scheduled for January and April 1989. The conclusions of the Expert Group should provide the basis for the substantive programmatic framework for the Decade, to be submitted to the 1989 session of the General Assembly. Reports on the first and second meeting of the Expert Group are available.

In addition to the two planning groups just described it is clear that initiatives are needed at the national level and in order to stimulate the establishment of national committees to this effect a letter suggesting possible modes of organization at the national level has been addressed to member States drawing their attention to the recommendations of General Assembly resolution 42/169. It was proposed that areas of activity for national committees might include:

– identification of hazard zones and hazard assessment;
– monitoring, prediction and warning;
– short-term protective measures and preparedness;
– long-term preventive measures;
– land use and risk management;
– public education and information.

Governments were asked to provide information on any action taken towards establishing a national committee or other official body for the Decade and many have already reported in the affirmative.

We are, of course, expecting national officials responsible for meteorological and hydrological services to join national committees together with representatives of other concerned disciplines and sectors, including, for instance, civil defence.

Another important body of participants in the Decade is the non-governmental sector. Relevant scientific and voluntary organizations have been contacted and so far most of them have replied positively.

I hope that this summary information on the preparations for the Decade has been of interest to you and UNDRO will, of course, continue to keep you informed, both through the World Meteorological Organization and other channels.

Hydrological Aspects of Extreme Floods and Droughts

JOSEF HLADNÝ and JOSEF BUCHTELE

Czech Hydrometeorological Institute, Prague, Czechoslovakia

Introduction

NATURAL extremes such as floods and droughts have accompanied mankind throughout its entire history. Witnesses to the destructive effects of floods and the negative consequences of droughts include ancient legends (the Deluge of the Bible, the seven lean years of Egypt), as well as artefacts coping with shortages of water (irrigation systems, and aqueducts supplying water to the ancient cities of the Orient). However, the paintings preserved in the caves of today's deserts, for example in southern Sahara, which depict one-time lively pastoral scenes in that region, bear out the fact that people's struggle with these natural phenomena was sometimes far from successful. Although the causes for this have varied, many of them exist, unfortunately, even today. And so the word 'disaster' has expressed almost the same meaning for a very long time – danger to people's lives and vital interests.

What are somewhat different, however, are the dimensions of the causes and consequences of extreme floods and droughts. For instance, research results prove clearly that evidence of the negative impacts of human activities can be found also outside the affected areas, with serious consequences on the continental or even global scale. The number of casualties and the extent of damages are increasing, also due to the other disasters that usually accompany extreme floods and droughts, such as famine, dehydration, epidemics, diseases, fires, etc. The results of the current extreme drought in the Sahel, or the recent catastrophic floods in Bangladesh and Sudan (Watanabe, 1989) are illustrative examples.

One of our common goals should be to reduce the danger foreseen by Sugawara (1978) in his outstanding WMO Bulletin article: ' . . . science and technology are effective for disasters of less than average magnitude, but are powerless for extraordinarily large events. Moreover, there is even the possibility of increasing the damage sustained in major disaster . . . ', because, ' . . . when a disaster is very large external aid cannot be expected. Then the will and ability of individuals to fight against disaster becomes the most important factor and I am very much afraid that this factor is currently growing progressively weaker. This will surely be the cause of a large disaster at some time in the future.'

Reducing the damage caused by floods and droughts and preventing loss of human life is based on gaining an increasingly better insight into the complex causes of these phenomena and on educating the people. The experience acquired so far indicates that the problems involved are of a broad, inter-disciplinary nature and, in many respects, unsolvable without international co-operation.

Current Knowledge of the Origination of Floods and Droughts and their Consequences

Studies dealing with floods and droughts undoubtedly rank at the top of the list of hydrological topics worldwide. They can be categorized into the following four groups:

(1) descriptions and evaluations of extreme phenomena, including analysis of their socio-economic consequences;
(2) information on the development of technological tools and operational systems;
(3) studies reflecting scientific achievements and new experimental approaches;
(4) philosophical and critical treatises which analyze the state-of-the-art in hydrology; although relatively small, this group usually presents sugges-tions, on an interdisciplinary basis, concerning the concepts and directions for further development.

The sensitivity of environment and national economy to the impact of floods and droughts is becoming ever more pronounced. This is a generally known reality of our times, and is being indirectly reflected in these studies, especially those of the 'Quo vadis' type. The causes for this are not unclear. Let us mention some of the chief ones:

– increased density of population in inundation areas;
– the process of urbanization and the higher density of housing in such regions;
– people's markedly changed way of life in respect to nature and natural resources (agriculture, mineral extraction and processing, etc.).

Flood damage that formerly was caused by, say, discharges having a return period of 100 years, can today be the result of a 20-year maximum discharge (Framji, Garg, 1976). Similarly, drought having the same interval of

9

recurrence may today affect populations several times larger in size than before.

In this context, the question will occasionally arise whether hazardous events such as floods and droughts can be prevented. The experiments with enhancing or artificially generating precipitation – some of which have been conducted with WMO support – have failed, and protection against the harmful effects of hail by way of shooting into storm clouds is very spot-related and can hardly be used routinely for large areas. For the time being, far greater hope for people and national economies lies in the defensive strategy (i.e. preventive construction, warnings and forecasts) of fighting floods and droughts rather than directly eliminating the actual, natural causes of their generation.

Linkages between floods/droughts and global processes

The question of how long in advance it is theoretically possible to forecast floods and droughts is not answered easily; the well known prognostic rule must always be employed: 'Forecasting on a scientific basis is impossible without understanding the causes of the phenomena to be forecast.'

Several fundamental moving forces can be traced, which generate floods and droughts. Our knowledge suggests the effect of the following basic factors:

– processes taking place outside and inside the Earth, manifested through geological, geophysical and oceanographic phenomena. The Earth's gravitational and inertial movements are often regarded as fundamental in this respect (Street-Perrot et al., 1983; Bucha, 1988).
– extraterrestrial influences, with the leading role played by solar activity and transmission of solar energy (Bhalme and Jadhav, 1984).

These processes produce combined effects on the world's system comprised of the atmosphere, hydrosphere, continents and the biosphere, thereby triggering complex interactions and feedbacks between its components. This is also the way in which the state of the atmosphere is created at every given moment. Weather is then defined as the observed instantaneous state of meteorological elements and phenomena existing in close and intricate inter-dependence. Weather averaged over a longer term in a given area is then viewed as climate. These atmospheric processes result in precipitation which is the immediate carrier of 'signals of latent drought or flood'. Determining the theoretical limit of the predictability of the causal meteorological phenomena will therefore provide an important basic item of information needed to form an opinion on the predictability of floods and droughts.

Meteorologial limits of predictability

This problem has been dealt with by many meteorologists, among them Smagorinski (1969), Mintz (1964), Charney (1966) and Pechala (1978). Using

10

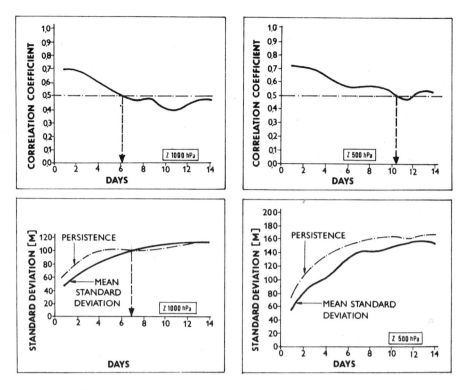

FIGURE 1 *Limits of meteorological predictability, after Miyakoda* et al. *(1972)*

different models of the atmosphere they found the limit of predictability to range from five to 22 days. Also Miyakoda (1972) compared the effectiveness of forecasts obtained on the basis of a nine-level atmospheric model with the persistence approach by means of a coefficient of correlation having an arbitrary limit of $r = 0.5$. The results, shown in Figure 1, suggest that in this case the interval of predictability is about one week for surface fields and about two weeks for free-atmosphere fields. In practice, however, forecasting relies on values below the lower boundary of this interval because the quantitative precipitation forecasts (QPF) appear to be one of the least predictable meteorological elements.

WMO Regional Meteorological Centres therefore usually issue QPF for only one to three days. But, in the event of a flood or in rainless periods, the situation and needs are very different. The specific features of the two phenomena, such as their duration, the imminent threat they pose and the consequences they produce, necessitate the use of different methods of prediction.

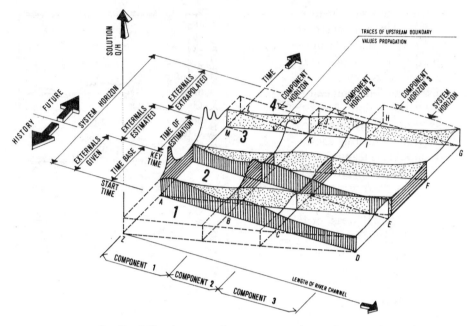

FIGURE 2 *Predictability in space-time representation, after Zezulák (1987)*

Causal meteorological factors of flood predictability

In the case of floods, all of the fundamental processes take a rapid course, developing in a relatively short time, which is why fighting their disastrous results entails urgency and, not very rarely, moments of surprise. The theoretical predictability of floods is usually limited by the predicability of:

– flood discharges and water levels in the river system in which the flood is moving (Zezulák, 1987), see Figure 2;
– the factor that has triggered the flood wave in the first place.

The diversity of river systems causes the predictability for individual basins to differ greatly, even for basins of the same size. It would certainly depend also on the type of flood in question.

Rainfall flood predictability, i.e. the lead time available for preparing forecasts, disseminating information, making decisions and adopting flood control measures, changes with the watershed area and other physico-geographical properties of basins. The predictability of rainfall of frontal type is greater than that of isolated rainstorms arising inside air masses of the same origin and unstable due to local heating.

Flash floods in rapid-response basins require a warning system whose most important components include a network of rain-gauging stations and a data collection system. There exists no exactly defined borderline between flash floods and floods caused by regional rainfall, but the predictability of flash floods largely depends on the information available on the causal rainfall, while for the latter type of flood additional data are essential.

Floods caused by snow and ice melting are typical of middle latitudes – the causal factor, air temperature, usually exhibits a higher degree of predictability than precipitation. Also, the flood's course is slower since even the fastest snowmelt merely approaches moderate rainfall. An exception is flooding produced by melting snow in föhn situations accompanied by rain. Seasonality is a prominent feature of this type of flood, making flood control easier but, at the same time, a greater accuracy of forecasts is required due to the large volume of flood waves.

Floods caused by strong wind or tide occur when strong wind or tide drives sea waves to coastal areas, or by backwater surging in estuaries of water streams. The waves produced by tropical cyclones are hurled against the coast with a destructive force. Sea waves whipped up by Atlantic hurricanes and Pacific typhoons are often compared to 'tsunami' – huge waves caused by the movement of the Earth's crust – which are not a hydrometeorological phenomenon.

Debris flows occur mostly in summer in the wake of rainstorms. In late spring, they appear due to the intensive snowmelt in the mountains, while in summer because of the rapid thawing of glaciers. Two approaches to protection are viable: the building of appropriate preventive facilities, and (automatic) warning stations. Long-term prevention includes measures that can help minimize the erosion produced by agricultural activities, urbanization processes, etc.

Besides predictability, the effect of lead time on hydrometeorological forecasting is influenced by the following factors:

– loss of time incurred in the operation of the reporting network and/or in forecasting procedures;
– transmission of warnings and forecasts to the users, and the decision-makers' readiness to take purposeful action;
– knowledge, experience and training of personnel.

The experience acquired in the training of professionals through the use of hydrological and water-management computer games played under the conditions of a simulated flood may help pave a road towards more effective warnings (Blažek and Kubát, 1989).

Meteorological factors and predictability of droughts

In contrast to flood, the term 'drought' can be defined with much greater difficulty. Meteorological drought is associated with fluctuations in atmospheric circulation. Under certain conditions, dry continental air stays over a given

J. HLADNÝ AND J. BUCHTELE

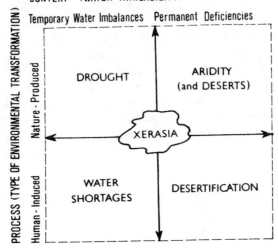

FIGURE 3 *Summary of each category of xeric regime, after Yevjevich, da Cunha and Vlachos (1983)*

area, anticyclones are created and hinder the movement of air masses, which prevents the wet ocean air from penetrating this area. Drought is manifest in long-term deficiency of precipitation, usually accompanied by high air temperatures. The decline in soil water-content then causes slower growth and even wilting of vegetation. This situation is regarded as agricultural drought. The critical diminishing of water storage – surface water (rivers, reservoirs, etc.), subsurface water (unsaturated zone) and ground water (saturated zone) – is listed in the hydrological drought category. Thus, hydrological drought is the effect of reduced water storages on certain non-hydrological systems or processes.

The nature of drought renders it in one of the categories of the so called 'dryness' or 'xerasia'. From this aspect, one should differentiate between the following four types of drought regime: aridity (deserts), desertification, water shortage and drought, as illustrated in Figure 3 – see Yevjevich, da Cunha and Vlachos (1983). One of the important characteristics of drought is the pattern in which it occurs, and we can therefore discern seasonal drought and random drought.

In the event of a drought already in progress, the persistence of low flows makes it possible to issue forecasts with longer lead times than in periods of average streamflows. Basically, this persistence depends on the conditions that govern reduction of water storages – river channels, the saturated zone, etc. – and is therefore site specific. Persistence is also related to antecedent conditions, as shown schematically in Figure 4.

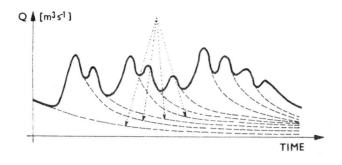

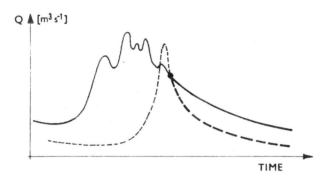

FIGURE 4 *Concepts of discharge computations for rainless periods. Top: Discharge as the summation of runoffs from different impulses; bottom: Variability of recession curve with different antecedent conditions*

Low flow predictability has recently been dealt with in more detail by Zhidikov and Mukhin (1987). When, however, the occurrence of drought is to be forecast with a great advance, the efficiency of such a forecast cannot be expected to be very successful. Figure 5 is an example derived from data on Indian rivers. This figure provides evidence that the distribution of drought situations, as represented by white circles, does not exhibit any coherence with the sunspot cycles. The problems encountered by attempts to identify the governing physical mechanisms that would serve as justification for endorsing the empirical relations used, constitute a difficult task in its own right. Moreover, the seemingly interesting result may often be distorted by the 'pseudo-cycle' effect. The traps entailed in this method of long-term drought prognosing, based on the identification of cycles, are revealed conclusively by Klemeš (1987).

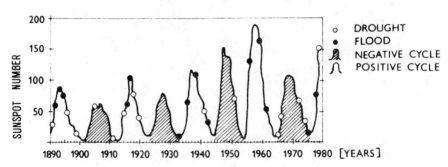

FIGURE 5 *Variations in mean annual sunspot number from 1891 to 1980, after Bhalme and Jadhav (1983)*

State-of-the-Art and Technological Possibilities of Monitoring

When assessing the quality and duration of observations, more often than not the conclusion may be drawn that the measurements are not quite satisfactory, particularly as regards the evaluation of extreme values and their probabilities. One of the frequent complaints concerns the inhomogeneity of the data. Although not always justified, serious problems turn up in connection with the non-stationarity of the runoff process, whether artificial or natural.

Accuracy of the basic elements

Data on discharges are, unfortunately, the least reliable of extreme phenomena. The reasons for this often include the inaccessibility of sites for hydrometric work and limited time for measurement during flood, the unstable gauging sites, variability of the discharge-rating curves, see Figure 6 (Clark, 1982), data inhomogeneity due to unregistered water abstraction in drought periods, frozen float-gauges, etc. An error of 40% of the maximum discharge value for large inundations during floods is not extraordinary. The situation is very similar for wide, unstable river beds during minimum discharges.

For the normal density of rain-gauging stations an error of about tens of percent in mean areal precipitation in smaller basins is not unusual, as cited by Clark (1982) and Eagleson (1967), see Figure 7. This can considerably influence also the accuracy and method of computing the design variables such as the probable maximum flood (PMF) and the probable maximum precipitation (PMP), and also the reliability of rainfall-runoff simulations. For example, the sensitivity analysis carried out by Burnash (1983) using the Sacramento model illustrates very clearly the paramount importance of precipitation data as compared with the accuracy of any of the model's parameters. Experiments with the same model and evaluations conducted by

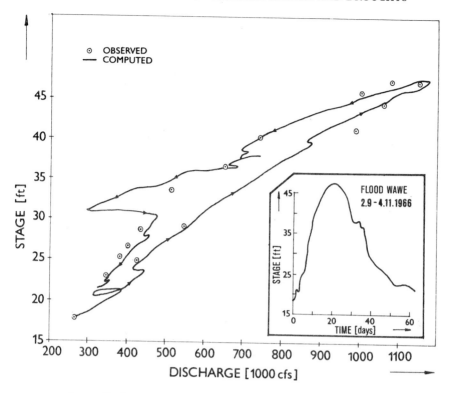

FIGURE 6 *Stage-discharge relations for the Mississippi river, Tarbert Landing, after Clark (1982)*

Němec-Sohaake (1982) indicate that the simulation of runoff for arid areas is particularly sensitive to the accuracy of inputs/outputs.

Equipment and organization

A short lead time results (see Figure 8) in efforts for its maximum utilization or prolongation on the basis of the data obtained from:

- a network of automatic rain gauges (Clark et al., 1983);
- a national network of meteorological radars furnished with computer and digital data processing system;
- meteorological satellites – geostationary and polar-orbit. This method enables to combine an almost continuous lower-resolution monitoring with more detailed scanning performed only a few times a day;
- quantitative precipitation forecasts (QPF) which may be produced using the data obtained by means of the above methods.

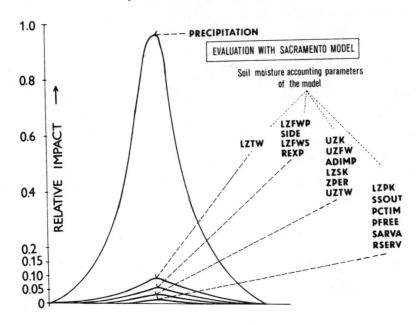

FIGURE 7 *Incremental effect of 10% changes in basic input of parameter values, after Burnash (1983)*

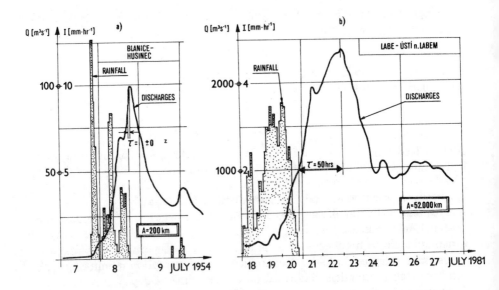

FIGURE 8 *Floods in basins of different sizes and corresponding lead times*

The hardware needed for data processing also includes telecommunication systems and computers. The recently introduced distributed systems and microcomputers are usually more flexible, accessible and reliable than mainframe computers and centralized systems.

The different forms of organizations at the national level as well as the diverse natural and technical conditions, give rise to the diverse approaches to the utilization of the above means. What, however, is an apparent trend is the endeavours to combine the data obtained from automatic rain-gauging stations, radars and satellites:

– automatic rain-gauging stations offer a reliable and instantaneous picture of the situation inside the basin, as well as direct inputs to hydrological models and data for calibrating radar reflectivity in real time;
– radar permits an insight into the dynamics and movement of rainstorms within a broader framework than a single basin;
– meteorological satellites provide data on regional precipitation structures, cyclones, etc. Satellite data also constitute a substantial part of the inputs to the QPF which, when combined with hydrological systems, may create a complex stochastic-dynamic hydrometeorological model (Georgakakos and Hudlow, 1983), see Figure 9.

Satellite data can also be helpful when examining various other phenomena:

– state and development of snow deposits;
– changes in the soil water content;
– characteristics of basin, (inundation areas, vegetation, ice phenomena, water quality, groundwater, etc.).

Trends in the Development of Methods for Hydrological Analysis and Forecasting

Trends in the application of model approaches

Until quite recently, the development of deterministic models has mostly been associated with forecasting needs, especially with respect to floods. Stochastic models have been considered to be the means for time series analysis and evaluation of data for design and planning. In connection with the existing and expected changes in the runoff process, caused by climatic variability and anthropogenic activities, there are efforts to investigate and assess these changes through deterministic approaches, namely rainfall-runoff and water-balance models. These tendencies are obviously associated with the understanding of the limitations entailed in the statistical methods of time series analysis, as well as with the fact that hydrology is being increasingly viewed as a geophysical science in which greater emphasis is placed on the

19

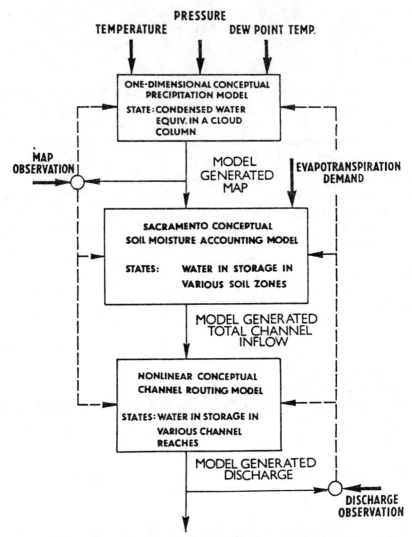

FIGURE 9 *Schematic representation of the stochastic-dynamic model of Georgakakos and Bras (Georgakakos 1986) ――― Model Generated Quantity; ▬▬ Exogenous Input or Output; _ _ _ Feedback Coupling by the Filter*

understanding of the physical processes that influence and form the runoff process (Klemeš, 1987).

When applying models to hydrological forecasting, interest oscillates between simple techniques, such as black-box models, and conceptual, complex models like the NWS (1972).

The efforts towards a simple structure of forecasting procedures are related with:

– attempts to analyze noises within the forecasting interval;

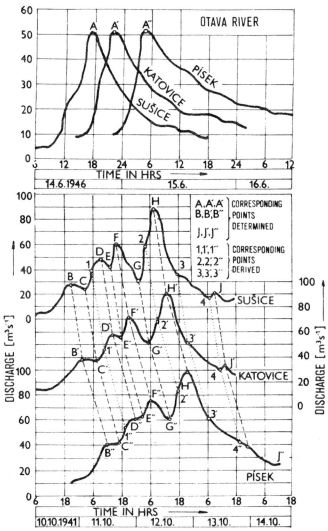

FIGURE 10 *Corresponding characteristic points for empirical relation (travel time forecasts)*

– tendencies to judge changes in runoff by local peculiarities and/or empirical relations, see Figure 10.

The strive for the structured model reflects an interest to take into account the areal non-uniformity of inputs (precipitation, changes in evapotranspiration and soil water content), hence, distributed models. Remote sensing techniques can be expected to become increasingly used as sources of data for such models (Deutsch *et al.*, 1981).

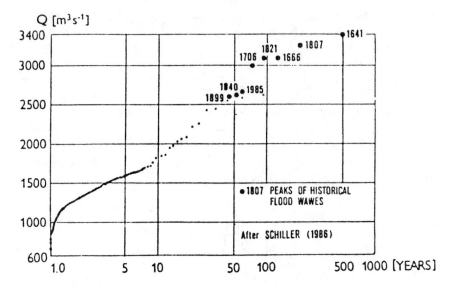

FIGURE 11 *The significance of historical floods for estimating probability density function parameters*

Problems of frequency analysis application

Despite having been used for decades, frequency analysis is the target of both research and objections, especially as regards the questions of probability distribution functions.

The usually available series (30- to 50-year series) do not permit a sufficiently reliable examination of asymmetry. This frequently necessitates the use of additional information for determining the parameters of the probability density function (comparison with other basins, regional analysis, incorporation of historical floods, etc., see Figure 11).

Regional analysis focused on the elimination of random errors in the estimate of the coefficient of asymmetry (Cs) should obviously include the ascertaining how this parameter depends on the flow forming factors such as the size and shape of the basin, some of the overall water yield indices, the basin's gradient, density of the river network, etc.

As for estimating extraordinary floods, worth mentioning are the research studies that concern two-component distributions which should be able to reproduce flood sets with a much higher fidelity even when exhibiting a great degree of asymmetry on which the estimate of extreme values strongly depends.

Still more problematic is the use of frequency analysis, for evaluating the flood hydrograph (or the discharge volume), since the chance to correctly estimate the parameters of a multi-dimensional distribution is markedly lower.

The calculations of design floods are therefore often based on precipitation data. This approach gives rise to uncertainties concerning both the estimate of the parameters that determine the runoff for extreme, i.e. usually unobserved conditions, as well as the assumptions about the areal homogeneity of runoff.

The statistical analysis of drought is even more difficult because of the occurrence of multi-year droughts. Regional analysis is not bringing about the information gain required since drought events in a homogeneous region usually occur at the same period for each stream. Starting points may include the approaches suggested by Klemeš (1987) in which a leading part will be played in water balance based on evapotranspiration as well as evaluation of soil moisture and groundwater inventories.

Evaluation of flooding phenomena in estuaries

Both simple relations and complex hydrodynamic computations and procedures known in oceanography are used as models for the prediction of estuary flooding which is the combined result of tidal phenomena, floods and surges caused mostly by wind. Both of these approaches need the data which are missing, particularly in developing countries where the problems are the most difficult due to the high density of population in fertile inundation areas. Wind directions, derived from satellite pictures in coastal regions, can be useful as input and effective means for lead time prolongation and for increasing forecast accuracy.

Conclusion and Recommended Measures

The above-outlined analysis of the hydrological aspects of floods and droughts is conducive to some general conclusions on the basis of which specific measures can be recommended.

Conclusions

(a) There are two types of hydrological support to protect against the devastating effects of floods and droughts:

– design values for protective facilities and measures (as for instance construction of reservoirs, plans for flood control, criteria for water management in drought periods);
– warning and forecasting systems.

The preceding point is associated with financial and material inputs and legislative steps, and their development will be relatively slower than that of warning and forecasting systems. When no protective structures exist, these systems constitute the only means of warning, and consequently adopting measures for the protection of human lives and property.

(b) The equipment and methodologies needed have undergone considerable development, and can assure better results than those achieved throughout the world so far. However, hydrological services and their networks are not equipped equally, which is the primary reason for this unsatisfactory situation, particularly in those parts of the world in which the detrimental effects of flood and droughts are the gravest.

(c) The growth of population and mankind's socio-economic development are generating new problems, such as desertification, deforestation, a rising number of flash floods in urbanized areas (Huff, 1986), agrotechnical procedures resulting in the erosion and compacting of soil, etc. Due to these factors, floods and droughts are becoming more devastating.

(d) The key to a longer lead time of hydrological forecasts is a deeper insight into the global processes. This approach should also help to:

– solve the problems of process teleconnection which may cause floods and droughts, like the El Niño effect;
– identify the temporal and spatial changes in the causes of flood and drought occurrence, as related to the expected climatic changes due to the higher levels of greenhouse gases in the atmosphere;
– evaluate influence of human activity on flood and drought occurrence.

Outline of feasible measures

Operational activities

Create conditions for a wider use of the means that exist in the framework of the WMO World Weather Watch (WWW) to meet hydrological needs:

– the use of telecommunication systems and means for the transmission of hydrological data, products of meteorological analysis and forecasts, including radar data. The aim is to set the ground for timely warning – especially in those countries that face the greatest threat of such phenomena as cyclones and surges and that do not as yet possess adequate technical or professional prerequisites for their own warning systems, i.e. primarily in subtropical and tropical zones;
– the endeavour for a wider utilization of satellite data in the evaluation of such phenomena as inundation areas, soil water content, groundwater occurrence, snow deposits and ice events, perhaps in co-operation with some other agencies (FAO, UNEP). Some of these observations are preconditioned by the launching of satellites carrying suitable equipment to suitable orbits;
– plan co-ordination and compatibility of the information provided by radar networks for exchanging data on international river basins.

Methodology, education and training

– encourage the development of new methods of measuring hydrological elements (direct registration of discharges, water levels in coastal areas,

snow deposits in mountainous and forested terrains, etc.);
- in co-operation with some other agencies, such as UNESCO and IAHS, increase efforts to gain new knowledge on the runoff process, hydrological models and hydrological cycle on the global scale, particularly through the atmosphere global circulation models. Strive for fruitful co-operation in other projects under way, concerning mainly the evaluation of evapotrans-atmospheric global circulation models. Strive for fruitful co-operation in conjunction with research going on within the WMO World Climate Research Programme (WRCP) and the WMO Typhoon Operational Experiment (TOPEX), etc.;
- intensify co-operation that helps utilize the knowledge on climate trends, the variability and changes in the assessment of the runoff regime undertaken within the WMO World Climate Programme–Water (WCP-Water);
- within the WMO Hydrological Operational Multi-purpose Sub-programme (HOMS), generate impulses toward the preparation of components and sequences focused on flood and drought evaluation;
- prepare teaching materials for the education of population in the sphere of flood and drought fighting. These should explain the negative impacts of anthropogenic activities on the runoff extremes on the global, continental and local scales – such as deforestation, erosion, desertification;
- organize international training seminars for the experts involved, in simulated situations, using, for example, training equipment systems.

Most of the problems that are to be coped with on the international level should meet with a corresponding response in national programs:

- continuous improvement of monitoring networks and their resistivity to effects of extreme situations (e.g. power and telecommunications failure);
- search for organizational forms that will support collaboration between hydrological and meteorological services;
- evaluation of all extreme flood and drought events;
- storage of data on extreme events in a unified form;
- establishment of a system for continual exchange of information, especially in international river basins;
- participation in the organization of educational activities.

REFERENCES

BHALME, H.N. and JADHAV, S.K. (1984) Double (hale) sunspot cycle and floods and drought in India. *Weather* **39**: 112–116

BLAŽEK, V. and KUBAT, J. (1989) Control training of Water Resources Systems under Simulated Flood Conditions. In: *Hydrology of Disasters* James and James, London, pp. 52–61

BUCHA, V. (1988) Influence of solar activity on atmospheric circulation types. *Ann Geophys* **6**: No 5

BURNASH, R.J.C., FERRAL, R.L. and McGUIRE, R.A. (1973) A generalized streamflow simulation system. Conceptual modelling for digital computers. NWS and California Dept. of Water Resources

BURNASH, R.J.C. (1983) Design considerations for an operational real-time hydrologic data system for small computer systems. WMO/NOAA techn. conference on mitigation of natural hazards through real-time data collection system and hydrological forecasting, Sacramento, California, Sept. 1983

CHARNEY, J.G. (1966) The feasibility of a global observation and analysis experiment. *Bull Am Meteorol Soc* **47**: 200–220

CLARK, R.A., BURNASH, R.J.C. and BARTFELD, I. (1983) ALERT: A National Weather Service Program for a locally-operated, real-time hydrologic telemetry and warning system. WMO/NOAA techn. conference on mitigation on natural hazards. Sacramento, California, Sept. 1983

CLARK, R.A. (1982) Scientific foundations of hydrological forecasting including optimization of operational service. Tech. note of US NWS Washington DC

DEUTSCH, M., WIESNET, D.R. and RANGO, A. (EDS) (1981) Satellite hydrology, *Proceedings of the Pecora Conference*, Am. Water Res. Assoc., Minneapolis, Minnesota

FRAMJI, K.K. and GARG, B.C. (1976) Flood control in the world, ICID, New Delhi

GEORGAKAKOS, K.P. (1986) A generalized stochastic hydrometeorological model for flood and flash-flood forecasting. *Water Resources Res* **22**: 2083–2095

GEORGAKAKOS, K.P. and HUDLOW, M.D. (1983) Quantitative precipitation forecast techniques for use in hydrologic forecasting. WMO/NOAA tech. conference on mitigation of natural hazards, Sacramento, California, Sept. 1983

HUFF, F.A. (1986) Urban hydrometeorology review. *Bull Am Meteorol Soc* **67**: 703–712

KLEMEŠ, V. (1987) Drought prediction: a hydrological perspective. In: *Planning for drought, toward a reduction of societal vulnerability*. Wilhite, D.A. *et al.*, (Eds) Westview Press, Boulder and London, UNEP

KLEMEŠ, V. (1985) Sensitivity of water resources systems to climate variations, WMO Report No. WCP-98

MIYAKODA, K., HEMBREE, G.D., STRICKLER, R.F. and SHULMAN, I. (1972) Cumulative results of extended forecast experiments. I. Model performance for winter cases. *Monthly Weather Rev* **100**: 836–855

MINTZ, Y. (1964) Very long term global integration of the primitive equation of atmospheric motion. Technical notes No. 66, Geneva WMO, pp. 144–167

NĚMEC, J. and SCHAAKE, J. (1982) Sensitivity of water resources systems to climate variations. *Hydrol Sci J* **27**: 327–343

NWS – OFFICE OF HYDROLOGY (1972) National Weather Service River Forecast Procedures – NOAA Technical Memorandum NWS HYDRO-14, NWS/ NOAA, US Dept. of Commerce, Silver Spring, Maryland

PECHALA, F. (1978) Predictability of meteorological phenomena. *Metorolog Zprávy* **31**:

SMAGORINSKI, J. (1969) Problems and promises of deterministic extended range forecasting. *Bull Am Meteorol Soc* **50**: 286–311

STREET-PERROTT, A., BERAN, M. and RADCLIFFE, R. (EDS) (1983) *Variations in the global water balance.* D. Reidel, Dordrecht

SUGAWARA, M. (1978) On natural disasters – some thoughts of a Japanese. *Bull WMO* IV

WATANABE, M. (1989) Some problems in flood disaster prevention in developing countries. In: *Hydrology of Disasters.* James and James, London

YEVHEVICH, V., DA CUNHA, L. and VLACHOS, E. (1983) *Coping with droughts* Water Resources Publ., Chelsea

ZEZULÁK, J. (1987) Hydrodynamic methods of system modelling for the needs of operational hydrology. Research report of the College of Agriculture, Prague

ZHIDIKOV, A.P. and MUKHIN, V.M. (1987) Technical report on methods of long range forecasting of low floods and droughts. WMO, Commission for Hydrology

Hydrological Systems Applied to the Management of Drought and Floods

—

PETER D. WALSH

North West Water Authority, Warrington, United Kingdom[†]

Introduction

THE PROVISION of hydrological assessments and forecasts is essential to support day to day river and water supply management functions including effluent treatment, pollution control, fisheries, land drainage and flood defence. Planning for construction projects also requires hydrological studies, particularly of extreme events to determine appropriate levels of protection.

Hydrological support is provided by a specialist team, which over the last 14 years has built up a repertoire of analytical methods and models. Considerable emphasis has been given to the development of standard methods. Most of the analytical work and data handling is by computer based 'decision-support' systems and these standard methods are therefore well defined and documented. Improvements and developments are mainly at the fringes of the systems and relate to enhancements in information generated by the core analysis or to its presentation and to ease of use by development into 'expert-systems'.

One of the greatest problems facing the support hydrologists is in presenting clearly and precisely the key issues on which executive decisions must be made. As these will affect the survival and economic livelihood of large numbers of people the hydrologist must not confuse the process by offering too many alternative scenarios of the likely developments for floods over the next few hours or days, or of droughts over coming weeks and months. The increasing power of computers, the profusion of statistical techniques for water resources analysis and the enormous number of different flow forecasting models now available mean that it is far too easy to ask and answer 'what if'

†Now with National Rivers Authority, North West Region, Warrington, United Kingdom.

questions for a wide range of assumptions using a variety of techniques and models. A major challenge for the hydrologist is therefore to select an approach which is appropriate to the situation in which it will be applied, to the data available and to the people who will use the information.

Hydrological problems of extreme floods and droughts may be examined from a number of viewpoints. Time is an unavoidable dimension, which may influence the approach taken in a number of ways:

(a) For example, the professional hydrologist is faced with two different technical problems, depending upon the time of occurrence of the extreme event:
   - the assessment of the severity of past and of current events;
   - the forecasting of the direction and magnitude of an event in the immediate future.
(b) The timescale of hydrological extremes may range from a few minutes to many months, even years:
   - floods are of comparatively short duration;
   - droughts may sometimes be of indeterminate length.
(c) The need to do work in real-time or in hindsight:
   - forecasting needs to be done in real time;
   - appraisal of past events or design of new schemes and engineering works are off-line activities.
(d) The hydrologist must also address problems of timeliness:
   - when forecasting floods time is of the essence;
   - drought forecasts need not be instantaneous.

Discussion of two major hydrological systems, their development and underlying philosophy illustrates the importance of integrating the analytical techniques with the human and organisational aspects, as well as with data processing and communications. Availability of data, frequency of measurement, processing effort and its timeliness, are other key factors for successful implementation and operation of hydrological support services for the management of extremes. The importance of meteorological products and their application to a diverse range of hydrological services has been described elsewhere (Walsh *et al.* 1987).

The discussion of the role of hydrology in the management of water resources systems describes the approach during periods of shortage. The principles inherent in the methods of analysis which lead to effective allocation of a scarce commodity (water) to meet society's needs are centred on the analysis of risk. Particular emphasis is given to the use of proxy data (e.g. from nearby catchments) and to detailed assessment of future risk levels.

The description of a flood forecasting and warning scheme contrasts not only the other hydrological extreme but also the significance of adequate and early warnings before the event. In many cases the increased precision of a forecast is of little importance if it delays the warning.

Drought Management

Many of the hydrological procedures used for controlling the operation of water resources have their origins in the methods applied to planning and design studies. These have been developed to derive both operating policies and decision support applications. In this way operational staff and managers have become increasingly familiar with, and have learnt to trust, the basis on which advice is produced for water resource systems.

Conjunctive use (Law, 1965) enables the hydrologically most appropriate sources to be used in the strategic allocation of sources ensuring maximum use of the cheaper (usually reservoir) water. Under normal circumstances, the rate of abstraction from reservoirs is set by reference to a control curve relating storage volumes and time of year with risk and rate of withdrawal. These curves are derived by assessing cumulative minimum runoff volumes from historic flow records (Walsh, 1971) A suite of computer programs (Pearson and Walsh, 1981) performs all the analysis, usually on historical records of flow. Where data are scarce or inadequate, a regional approach is used based on data from nearby catchments and estimates of long-term average runoff (Pearson and Walsh, 1982). The underlying approach to these methods is that of the mass-curve (Rippl, 1882/3) and its extension into critical period techniques using non-sequential cumulative runoff volumes (see McMahon and Mein, 1978) on a seasonal (time of year) basis.

A recent review (Walsh et al., 1988) of the development and application of these hydrological techniques shows how these have become the cornerstone of operating policies to achieve optimum economic levels at predetermined levels of risk. In Autumn of the previous year allocation studies are performed for the demand level anticipated for the following Spring onwards. Economic appraisal, including the use of dynamic programming for resource systems with large and expensive pumped sources, determine optimum policies (Walker et al., 1989) and lead to the production of a strategic operating plan for the following year. This is produced jointly by hydrologists and operational managers.

A natural extension, of this close working relationship and of the hydrology inherent in the overall approach to water resources design and operation, is into systems and methods for drought management, which are utilized when it is necessary to modify normal operating procedures.

The upper control curve line in Figure 1 shows that in 'State A' it is safe to take more than the reliable yield of the reservoir without jeopardizing the reliability of the source. It may be calculated for any risk level (probability) or from worst historic events. Analysis to produce this curve and for the drought risk assessments described below are carried out at monthly time steps to restrict both the amount of computation and of data to be interpreted within manageable limits. However, simulation models which are used to verify control curves and estimate operating costs usually operate at a daily time step.

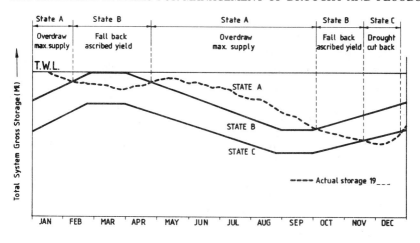

FIGURE 1 *Reservoir control curve*

The lower line in Figure 1 provides an indication that when the reservoir falls into 'State C', risks of failure should be specially assessed and further hydrological investigations should be performed to evaluate (a) the severity of the dry weather already experienced and (b) the risks to supply in the future for a range of withdrawal rates. These analyses use cumulative runoff volumes for the current and successive months for a range of probabilities and the minimum historic event to produce information such as that in Figure 2. This is repeated at each reservoir and the quantities to be supplied can be adjusted to equalize risk of failure across a number of reservoirs or other steps taken to implement emergency sources or restrict consumption by limiting use for some applications. The development of the decision support system (Walsh and Walker, 1989) that produces this information has taken place over the last five years during a major drought event (1984) and in response to potential droughts in other years. Experience has shown that the logistics of performing these appraisals can be considerable and that efficient reporting of runoff, rainfall and current reservoir contents and withdrawal rates is essential. Even when the data for many sources can be brought together in one place to meet a computer run deadline, there still remains the task of analyzing large volumes of computer results. There are considerable attractions from exploiting the potential of 'expert systems' (Palmer and Tull, 1987). Not only can the logistics of the exercise be considerably simplified, but the expert knowledge and approach established by the hydrologist becomes more widely available.

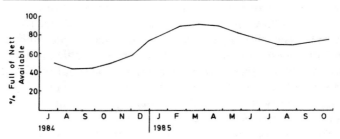

ASSUMPTIONS : Initial Storage as of 30/7/84   Draw to supply 58.0 Ml/d

Runoff Events starting 1st August

Compensation Water reduced

PREDICTED MINIMUM CONTENTS THIS YEAR

%. Full of Nett Available

	0	20	40	60	80	100	MONTH OF FAILURE
Minimum Historic	DEC						MARCH 85
2 %. Prob.	OCT						JULY 85
5 %. Prob.	OCT						1986
10 %. Prob.	OCT						
20 %. Prob.	OCT						
50 %. Prob.	AUG						

ESTIMATED YIELDS AND CRITICAL PERIODS

MINIMUM HISTORIC   40.9   Ml/d   over   14   months critical period.

5 %. PROBABILITY   56.4   Ml/d   over   26   months critical period

10 %. PROBABILITY *   60.5   Ml/d   over   26   months critical period

ESTIMATED PATTERN OF STORAGE WITH AVERAGE RUNOFF

FIGURE 2   *Assessment of water storage solution*

Flood Warning

A region wide scheme for flood warning using radio telemetry and computer systems was being implemented in the late 1970s as proposals for a prototype unmanned weather radar were under consideration. The proposals were brought together in the North West Weather Radar Project (Crowther, 1986) which was jointly funded by the Meteorological Office, other agencies, government departments and North West Water Authority. In 1989 the National Rivers Authority took over responsibility for flood warning from North West Water Authority.

The project embraced implementation and operational aspects of the radar hardware and its integration with river-flow and rainfall telemetry into a flood forecasting and warning system. It included the development of flow forecasting models (Reed, 1984) and appraisal of benefits (Anon, 1983). The results of all these studies have been published (Collinge and Kirby, 1987). The development of the operational flood warning scheme occurred concurrently (Noonan, 1986).

The National Rivers Authority North West Region's flood warning scheme is based on a telemetry system scanning 50 rivers and 20 rainfall measuring sites, together with the weather radar, automatically every 15 minutes. Radio and microwave links between outstations and the central telemetry computers also provide data and voice links between the flood warning centre and local offices. Telex and telephone are used to communicate with police and other, mainly local government, agencies. Duplication through a variety of communications methods (including private and public telephone, radio pagers, and electronic mail) ensure that communications are virtually secure.

The best possible hydrological forecast of flood levels and times is useless unless it can be disseminated promptly, in a form which is understood and acted upon quickly. Considerable attention has therefore been given to improving and reducing dissemination times and to ease of presentation within the flood forecasting system, rather than to increasing lead times through better hydrological modelling. Standard tabular reports on colour terminals and a flexible yet rapid screen presentation of hydrographs and rainfall rates allows rapid assimilation of current conditions. Simple models (Knowles, 1987) are run automatically and their predictions together with actual flows and cumulative rainfall volumes are monitored both by staff and the computer against predetermined thresholds. Exceedence of these thresholds causes the telemetry computer system to create 'alarm' messages which are passed by a 24-hour control centre to one of two Regional Duty Officers. These duty officers can interrogate the system with portable computer terminals from home if necessary, but would open up the flood forecast centre immediately if events exceeded 'standby' levels and the need to contact other agencies was likely. Other parts of the National Rivers Authority are advised at the same time but no messages are transmitted to the police until a further high threshold has been reached. At this level the police, whose responsibility it is to warn the public, become aware that flooding might occur. A 'warning' is only issued when forecasts indicate that the flood level will be exceeded.

Simple telex messages advise the police that flooding will occur at places that have been identified to them through schedules and maps already in their possession. In this way as much information as possible is prepared in anticipation of flooding leaving only the minimum of essential information to be transmitted during the flood event. All the organisations involved in responding to flood warnings and flood events have clearly defined roles and responsibilities. This ensures the minimum of delays both in transmission of warnings and in their response to aid the population in flooded areas. Regular meetings are held with police and local agencies to review lessons learnt during an event and to update warning schedules and arrangements. It is unusual for the same police officers to be involved in more than one flood event and for this reason the documentation and schedules must be clear and concise.

Conclusions

Whilst it is important to ensure that sound and technically correct methods of hydrological analysis and modelling are always used, these must be incorporated within a total systems approach that will be effective in application at times of stress. The total system must be designed to fit within the organisational and administrative structures. Unless this is achieved there will be communications difficulties, confused responsibilities and inadequate action to deal with a hydrological disaster on the ground. This paper has reported two approaches that have been developed to minimize the chance of mistakes occurring from these causes.

REFERENCES

ANON (1983) Report of the working group on national weather radar coverage. National Water Council – Meteorological Office Liaison Meeting, London

COLLINGE, V.K. and KIRBY, C. (1987) *Weather Radar and Flood Forecasting.* Wiley, Chichester

CROWTHER, L. (1986) The implementation of an integrated regional flood forecasting and warning service using weather radar. In: *Measurement of precipitation by radar,* Cost Project 72 Commission of the European Communities, Report EUR 10353, 74

KNOWLES, J.M. (1987) Flood forecasting hydrology in North West Water. In: *Weather Radar and Flood Forecasting.* Collinge, V.K. and Kirby, C. (Eds). John Wiley, Chichester

LAW, F. (1965) Integrated use of diverse resources. *J Inst Water Eng* **19**: 413

McMAHON, T.A. and MEIN, R.G. (1978) Reservoir capacity and yield. *Developments in Water Science,* No. 9. Elsevier, Amsterdam

NOONAN, G.A. (1986) An operational flood warning system. *J Inst Water Eng Sci* **40**: 437

PALMER, R.N. and TULL, R.M. (1987) Expert systems in drought management planning. *J Comp Civil Eng* **1**: 284

PEARSON, D. and WALSH, P.D. (1981) The implementation and application of a suite for the simulation of complex water resource systems in evaluation and planning studies. In: *Logistics and Benefits of Using Mathematical Models of Hydrologic and Water Resource Systems* (Proc. Pisa Symp. Oct. 1978) IIASA Proc Vol 13

PEARSON, D. and WALSH, P.D. (1982) The derivation and use of control curves for the regional allocation of water resources. In: *Optimal Allocation of Water Resources*. Lowing M.J. (Ed). IAHS Pub No 135

REED, D.W. (1984) A review of British flood forecasting practice. Report No. 90. Institute of Hydrology, Wallingford

RIPPL, W. (1882/3) The capacity of storage reservoirs for water supply. *Min Proc Inst Civil Eng* **LXXI**: 270

WALKER, S., WALSH, P.D. and WYATT, T. (1989) Derivation and application of medium term operating policies for the Northern Command zone system of North West Water (UK). In: Systems *Analysis for Water Resources Management: Closing the gap between theory and practice* Loucks, D.P. (Ed.) IAHS Publication No. 180, p. 53

WALSH, P.D. (1971) Designing control rules for the conjunctive use of impounding reservoirs. *J Inst Water Eng* **25**: 371

WALSH, P.D., NOONAN, G.A. and KNOWLES J.M. (1987) The application of meteorological information to water services and river basin management. World Meteorological Organisation symposium on education and training in meteorology with emphasis on the optimal use of meteorological information and products by all potential users. Shinfield Park, Reading, UK

WALSH, Y.D., WALKER, S. and PEARSON, D. (1988) Derivation of operating policies for surface water sources in North West Water. *J Inst Water Environ Man* **2**: 51

WALSH, P.D. and WALKER, S. (1989) Decision support systems as an aid in the operational management of multiple water source systems. In: Systems *Analysis for Water Resources Management: Closing the gap between theory and practice* Loucks, D.P. (Ed.) IAHS Publication No. 180, p. 103

Morphological Changes in the Swiss Alps Resulting from the 1987 Summer Storms

F. NAEF, W. HAEBERLI and M. JÄGGI

Laboratory of Hydraulics, Hydrology and Glaciology,
Swiss Federal Institute of Technology, Zurich, Switzerland

DURING the summer of 1987, several large floods in the Swiss Alps caused considerable damage, blocking major roads and train lines. The damage resulted largely from flood-driven erosion which brought about large scale morphological changes in several valleys and water courses. In the first part of this article, debris flows occurring in short and steep side valleys as the Varuna (Poschiavo), the Minstiger or Geren valley (Obergoms) are described. In the second part, details are given of the events in the 40 km long Reuss valley, which led to the blockage of the most important north-south connection over the Gotthard pass.

Debris Flow Events

Figure 1 shows cross sections of the lower part of the Varuna valley before and after the debris flow of 18 July 1987. During this event, the valley was eroded by more than 10 m over a length of 1,600 m, roughly 350,000 m³ of debris being washed out in the process. This material discharged into the main valley and dammed the Poschiavino river, causing it to change its course and flow through the main village of the valley, where it caused severe damage.

Similar events occurred a few weeks later, on 24 August 1987, in the upper Rhone valley, when the village of Münster was crossed by a debris flow. The front of this flow appeared to eye witnesses as 'a many metre high dark wall of blocks and trees, advancing with a speed comparable to that of a normally driven car, causing heavy vibrations as well as a noise like thunder'. An equally spectacular phenomenon occurred in the nearby Geren valley; it was less well observed however, because the flow came to a halt outside the inhabited area.

36

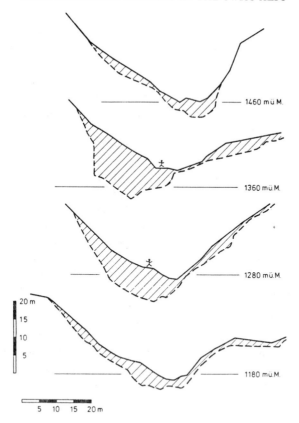

FIGURE 1   *Cross-sections of the Varuna valley before and after the debris flow of 18 July 1987. The hatched areas have been eroded by the debris flow*

Varuna Valley in Poschiavo

The last debris flow at Poschiavo of comparable size to the 1987 event occurred more than 150 years ago, in 1834. According to the chronicles from that time, the course and the consequences of the flow were similar to those of the 1987 event. In both instances the sides of the valley as well as the large and steep debris cones became destabilized. In the decades following 1834, more debris flows occurred and as a consequence the Varuna became known as one of the most dangerous mountain torrents in Switzerland.

Later, the Varuna became less active. Thanks to the construction of more than 50 check dams along its course, erosional and landslide debris could not be transported downstream. Instead it accumulated in the lower part of the Val Varuna, a V-shaped valley, 1600 m long with a slope of 40%.

In the 1950s, a gauging station was installed in the Varuna in connection with a hydroelectric power project. The records from 30 years of discharge measurements at this station disclose no evidence of flood discharge levels, nor even of a sharp rise in the discharge. This is surprising in view of the fact that the Varuna is a steep torrent, situated in an area where heavy thunderstorms are known to occur. Some of the runoff was apparently able to seep into the debris with the result that the discharge, flowing partly underground, was both delayed and attenuated. It appeared, therefore, that the Varuna had become tame. In reality, however, it would only take a particular combination of unfavourable factors to release an avalanche of accumulated debris down the valley, with disastrous consequences.

Just such a combination of factors occurred on 18 July 1987. After the late snowfalls in May and June, intense snowmelt could take place in the warm days preceding 18 July. The runoff from the heavy precipitation of 18 July 1987, which fell as rain even in the highest parts of the catchment because of high temperatures, topped therefore on already high discharge.

Meltwater from the small glacier on the east face of Piz Varuna (3,453 m a.s.l. and highest point of the catchment), forms a waterfall down a rock wall and onto a large and steep debris cone (Figure 2). At the head of this debris cone, at about 2,700 to 2,760 m a.s.l., the meltwater flowed under an arch in some old avalanche. After the debris-flow event, the arch had collapsed and ejected mud could be seen. The debris cone immediately beneath the avalanche snow patch was strongly eroded. Mud and water-saturated debris may have temporarily blocked the channel underneath the avalanche snow. A sudden release of this material then possibily triggered the debris flow, which in turn began to deeply erode the steep river bed all the way down the main valley (Figure 2).

Apart from a few other traces of debris flow in the debris cone in the upper part of the catchment, no other signs of extraordinary flows or exceptional discharge could be detected in the Varuna drainage basin. This is taken as evidence that the debris-flow activity was indeed primarily triggered by the peculiar processes occurring within the uppermost parts of the drainage basin.

Minstiger and Geren Valley

Comparable observations were made in the upper Rhône valley after the debris-flow occurrences of 24/25 August. Again heavy precipitation had fallen as rain up to high altitudes. The glaciers themselves did not show any noticeable traces of extraordinary events such as surge-type sliding movements, ice avalanches or outbursts of water pockets or ice-dammed lakes. However, steep slopes covered with loose and vegetation-free sediments from active debris cones and Little Ice Age moraines became destabilized in many places close to long lasting or even perennial snow patches and glacier margins. The most spectacular events occurred immediately beneath the Minstiger and Saas glaciers. In the Minstiger valley, the debris flow must be assumed to have

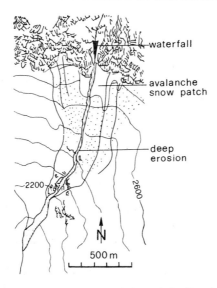

FIGURE 2 *Situation at the starting zone of the debris flow in the Varuna valley*

started in a steep rock gorge filled with morainic debris which had become ice free within the last few decades only. The triggering process may have been related to a displacement of the main subglacial drainage channel, and the corresponding formation of a waterfall into the steep proglacial rock gorge. The debris flow itself probably reached the main valley and the village of Münster without major depositional activity.

A voluminous debris flow occurred also in the Geren Valley. It too started in the forefield of a glacier, the small Saas glacier, and caused massive erosion of the loose sediments. The starting zone at the margin of a long lasting firn patch concerned a steep slope at about 2,540 m a.s.l. and exposed a partly overhanging, perennially frozen and obviously ice-rich moraine wall. Even several days after the event, debris and mud still continued to slide off this exposed permafrost zone, and the Saas river remained highly turbid in contrast to most other meltwater streams. Since the stability of the exposed and partly overhanging permafrost moraine was difficult to judge, the risk of a major slide and debris flow at the Saas glacier and in the Geren valley could not be excluded. The bottom of the Geren valley had therefore to be designated a high-risk zone and access was prohibited.

After the events

Today's situation in Münster and Poschiavo is fundamentally different. Loose material within the main starting zone of the large debris flow in the Minstiger valley, the rock gorge beneath the Minstiger glacier, has now been essentially

39

removed. The slopes of the relatively wide glacial valley remain virtually unchanged and a reoccurrence of the same event within the immediate future appears unlikely. However, in the starting zone of the debris flows within the Val Varuna, the potential for a repetition of the dangerous events exists; the steep, deeply weathered and therefore geologically weak slopes of the narrow fluvial valley have become destabilized, so that an acute hazard remains unless appropriate measures are completed.

Changes of the River Morphology in The Upper Reuss Valley (Canton URI)

The Reuss river has its spring north of the Gotthard pass, and its valley then crosses the canton of Uri in the South-North direction. At its mouth, where it enters the Lake of Lucerne, its catchment area is about 800 km^2. The heavy rainfall on 24 August that triggered the debris flows in Obergoms also affected the upper Reuss valley. On this day, it rained from the morning onwards and the discharge of the Reuss kept rising. One hour before midnight, the discharge matched the 10-year flood peak. But the water did not recede. In the next hour, more than 40 mm of rain fell on the already saturated upper part of the Reuss catchment basin. This resulted in a flood peak of short duration but of a size never recorded before. Widespread and heavy damage was the consequence. The most striking changes in riverbed morphology occurred between Göschenen and Amsteg, where the river follows an old glacial valley whose bottom is largely filled with loose material from moraines, rockfalls, and slope debris. While solid rock constitutes the river bed at many places, at others the bed has been formed by selective erosion: fine material has been washed out from the surface of the bed and only large boulders remain to form a natural armour layer.

During the extreme flood in August 1987, this armour layer was brought into motion over long stretches allowing bed erosion to take place. Dramatic channel deformation followed. Since the banks were not protected, but consisted mainly of residual boulders like those at the bed, bank erosion was an immediate consequence of bed erosion, and widening therefore predominated over bed incision. The active width of the bed more than doubled.

In addition to channel deformation, a horizontal oscillation took place, reminiscent of the meander formation normally associated with lowland rivers. Such a movement can be triggered by a natural curve or artificial change in the river's course. Intensive but localized attacks on the river banks occurred, followed by collapse of the slope and increased sediment input into the Reuss. Within a few hours the landscape changed dramatically (Figures 3 and 4).

In the decades prior to 1987, agricultural land use had spread very close to the active river bed, with only dry stone walls separating the fields from the river in many places. These walls were destroyed in the 1987 event and a

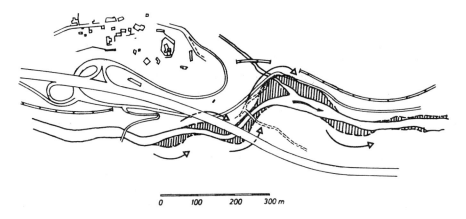

FIGURE 3 *Example of the bank erosion (hatched areas) in the River Reuss near Wassen due to the meandering of the river (indicated by the arrows) during the flood of 24/25 August 1987*

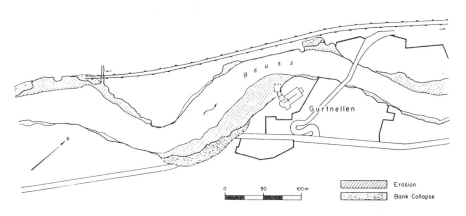

FIGURE 4 *Example of the bank erosion in Gurtnellen*

completely new bank alignment was the result. Since train lines and roads followed the old river bank over long distances, they were also affected. Their supports were embedded in the large boulders constituting the river banks, and it only needed the slightest movement of these boulders for the supports, and therefore the train lines and roads, to collapse.

The Gotthard railway and the old road over the Gotthard pass were built more than 100 years ago. In that time, they had never been seriously threatened by bank erosion. The 1987 event moved the residual blocks and activated the meanders. The resulting marked changes in the landscape have to be considered irreversible. The natural protection which had previously been provided by the large boulders for time periods of decades or possibly even centuries, has been altered and may be completely absent.

The description of this process contradicts the widespread opinion that a river channel is mainly formed by frequent floods of small magnitude and that extreme floods exert little influence only. River mechanics and river morphology tell us that unless certain threshold values are exceeded, no erosion or meandering takes place. Although standard threshold values cannot be applied in the case of steep mountain rivers, the observations made in the Reuss valleys show that similar relationships may well exist.

The events of August 1987 should be seen in relation to the general evolution of a postglacial river valley. It can be assumed that the conditions immediately after the retreat of the glaciers were unstable and that changes in topography were rapid and large. In time, selective erosion started to protect the river bed and the banks against further erosion. Only exceptional flood events which surpassed the previously mentioned threshold values were able to contribute to the further evolution of the valley. The evolution of this particular landscape is not a regular, continuous process, but takes place in abrupt steps (disaster theory). An event which has, by human standards, exceptional consequences is nothing but a short chapter in the geomorphological history of such a valley.

The Improbable Probabilities of Extreme Floods and Droughts

=

V. KLEMEŠ

National Hydrology Research Institute,
Department of the Environment, Saskatoon, Canada

Introduction

THE CURRENT approach to the estimation of probabilities of extreme floods and droughts is based on analyis of extremes in historic streamflow or precipitation records. The main weakness of the analysis is that it takes no account of the actual climatic, hydrological and other geophysical mechanisms that produced the observed extremes. Rather, it is based on arbitrary postulates of preconceived probabilistic mechanisms so that the results do not reflect what is likely to happen in nature, but what would happen if these postulates were correct. The crucial aspect, i.e. whether they actually are correct, is not addressed at all. The usual excuse for this is that there are not enough data, the physical processes are not understood well enough and the planners, engineers, managers, etc. cannot wait with their decisions until these problems are resolved; hence, to comply with the practitioners' requests for estimates of 100-year, 1,000-year, and (lately) even 1,000,000-year floods and droughts, these crude methods have to be used.

This attitude reflects a confusion between the requirements of current decision making and the need for improving the scientific basis for future decision making. In case of hydrologic extremes, the latter has been entirely sacrificed to the former – the above mentioned excuse has been repeated for at least 50 years during which time much progress on the real difficult issues could have been made but has not, so that the present-day planners are no better off than their grandfathers were.

V. KLEMEŠ

Probabilistic models of physical phenomena

Probabilistic modelling of physical phenomena faces one fundamental dilemma. On one hand, it is motivated by a realization that the causal structure (dynamic mechanism) of the given phenomenon cannot be explicitly described for one reason or another: it may be unknown, known but forbiddingly complex, unknowable, or it even may be considered nonexistent. On the other hand, the leading probabilists and statisticians seem to agree that a meaningful probabilistic (stochastic) model of a physical phenomenon cannot be constructed, unless the dynamics of this phenomenon are fairly well understood (some pronouncements to this effect by scientists like Wiener, Bartlett, Kendall and Stuart, and Box were quoted in Klemeš, 1978 and 1986).

The first pole of this dilemma leads, by default, to a replacement of the unknown dynamics with probabilistic assumptions which are compatible with the statistics derived from observations of the given phenomenon, but otherwise dictated by mathematical convenience and requirements of the available theoretical apparatus, sometimes even with full awareness of their conflict with reality. To give simple examples of the latter, one may mention the often invoked normality of a distribution of a phenomenon which has obvious physical bounds, or ergodicity in cases where, physically, only one realization of a 'process' can exist.

This route to probabilistic modelling is essentially a generalized curve fitting and may be acceptable (and often useful) for the reduction of data from observations, especially where they are required as inputs for various straightforward applications (e.g. mapping, smoothing, interpolation). A model of this nature may provide a good enought ad hoc description of a set of observations, but there is no reason to assume that this description adequately represents probabilities of states of a physical system on which these observations were made. This means that such a model may be completely useless and misleading for extrapolation beyond the range of observations.

It has been in recognition of this fact that statisticians have been pointing to the other pole of the dilemma, i.e. emphasizing the need for understanding the dynamics that has led to the observed states, as a necessary prerequisite for building a good probabilistic model. By 'good' model is meant one that has some measure of credibility when used for extrapolation, i.e. a model that can tell us more about the given phenomenon than the observations themselves reveal. It can be argued that only in such a case the term 'model' is justified; otherwise it is just an inflated name for an interpolation formula which, when used for extrapolation, may be an extremely unreliable and dangerous tool. In my opinion, this is the case with the overwhelming majority of approaches to the estimation of probabilities of extreme floods and droughts. In this regard, we have no hydrology of disasters; we only have a disaster of hydrology.

Probability in Hydrological Context

The general practice in contemporary hydrology is to regard the probability of a hydrological event as a limit of its relative frequency observed in the historical record. Two assumptions are crucial for this concept, in particular (1) that hydrological processes are stationary over time, and (2) that they are realizations of an ergodic stochastic process.

The first assumption (stationarity) implies that a hydrological phenomenon, e.g. streamflow at a given gauging station, fluctuates in time around a constant value in a statistically constant pattern. This further implies an assumption of a physical constancy of the mechanisms participating in the formation of the streamflow, from the regimes of precipitation and evaporation in the river basin, to geomorphological, pedological and other physical conditions. It is of course well known that this assumption is not true in general and that it diverges from reality with the length of the period considered. As the length of record increases and, with it, the nominal statistical significance of parameters derived from it, the validity of the stationarity assumption decreases. Thus, despite the preaching about the importance of long records, hydrologists are in fact more comfortable with short ones; the stationary hypothesis is much more easily defensible for, say, a 30-year record that it would be for a 300-year record. This leads to a paradox that the estimates of population parameters, population distributions and probabilities, the philosophical justification of which comes from the ideal of a very long record, makes sense only when based on a short record. Then, however, doubts arise about the validity of estimates of characteristics reflecting the long-term properties of the process. While it may be reasonable to assume, for instance, that the past 30 years may provide an adequate picture of the range of floods that one could expect in the immediate future since the physical conditions may not have changed much during that period, it is entirely different to say that, based on a distribution fitted to this 30-year record, a 1,000-year flood will be such and such.

In other words, making a statement about high extremes based on a statistical model fitted to a short record is similar to making a statement about extremes of a wavy curve based on the properties of extremes of a tangent representing it in a given point.

The second assumption (ergodicity) implies that the historic record of a hydrological phenomenon can be regarded as one of an infinite number of equally likely realizations of a stochastic process in which ensemble averages are equal to time averages. This is the most fundamental assumption on which rests the whole concept of 'probability of a hydrological phenomenon'. Without the assumption of an ensemble of realizations and the assumption that the incidence of possible events across the ensemble at a given instant is the same as was the incidence of events along the historic realization, the present concept of hydrological probability is unthinkable. And yet, from a geophysical point of view both these assumptions are not only arbitrary and

45

unrealistic; they deliberately make a mockery of reality, of the evolutionary character of the history of a geophysical process and of the uniqueness of this history. They practically reduce all the history to a noise which could have proceeded in the reverse direction as well as in the actual one or in any other rearrangement – and the same is implied for the future. They completely negate the fact that there were specific signals associated with specific events and that many of these signals, especially those associated with the very extreme events, physically cannot be repeated either at all or at least not with the same probability in every consecutive year or instant. They provide an excellent example where a mathematical concept developed to describe one physical situation (some phenomena in thermodynamics in this case) has been applied to a situation it does not fit for no other reason than mathematical convenience.

The leap of logic by which the instantaneous probabilities are equated with the historic frequencies of occurrence is nothing else but a dismissal of any meaning of the historic process: if anything that happened in the past can happen at any instant with the same likelihood, then the history provides no meaningful information.

We are facing perhaps the greatest paradox of probabilistic statements about hydrological phenomena. They claim to give information about the future – and they arrive at this information by first suppressing most of the information from the past, by denying any significance to the order of the past events.

The automatic identification of past frequencies with present probabilities is the greatest plague of contemporary statistical and stochastic hydrology. It has become so deeply ingrained that it prevents hydrologists from seeing the fundamental difference between the two concepts. It is often difficult to put across the fact that whereas a histogram of frequencies of given quantities (or the empirical return periods) can be constructed for any function whether it has been generated by deterministic or random mechanism, it can be interpreted as a probability distribution only in the latter case. A perfect sine wave will yield a histogram of frequencies which have not the slightest probabilistic meaning. *Ergo*, automatically to interpret past frequencies as present probabilities means *a priori* to deny the possibility of any signal in the geophysical history; this certainly is not science but sterile scholasticism.

The point then arises, why are these unreasonable assumptions made if it is obvious that probabilistic statements based on them may be grossly misleading, especially when they relate to physically extreme conditions where errors can have catastrophic consequences? The answer seems to be that they provide the only conceptual framework that makes is possible to make probabilistic statements, i.e. they must be used if the objective is to make such probabilistic statements.

It is like the well known joke where a policeman sees a drunk diligently searching for something under a street lamp.

'What are you looking for?' asks the policeman.

'I have lost my key.' says the man.

'And did you lose it here?'

'Well, no, but this is the only place where I can see anything'.

Extreme Floods

In contemporary hydrological practices the esimation of probabilities of extreme floods has become a rather trivial problem which requires no understanding of either hydrology, climatology, probability theory, statistics or any other relevant science. All that is necessary is to obtain the values of the maximum annual flows from the historic record of, say, 30 years, arrange them in order of their magnitude on an internal (0, 1), fit this empirical distribution of relative cumulative frequencies with a few simple S-shaped mathematical functions, select the 'best fit' by some standard curve fitting technique and read the exceedence 'probability' of any given flow off the upper tail of this curve as the value of the corresponding decimal fraction; alternatively, the value of this decimal fraction, say 0.001, is chosen and the corresponding value of the flow is designated as the 1,000-year flood.

All this is conveniently done by computers using one of the many available software packages for flood frequency analysis. The most demanding task in the exercise probably is to enter the historic maxima into the computer. However, the most difficult task, at least for this writer, is to believe that a number obtained in this standard manner has much to do with the real probability that a flood of the given magnitude could occur in any given year.

Some of the specific reasons following from the general criticism outlined in the preceding sections are given below. A short historic record may perhaps cover a period during which the climatic and hydrological regimes were close to stationary and the observed hydrological activity really reflected mostly the random fluctuations of energy within the system. However, without considering the types of energies taking part in these 'normal' perturbations of some underlying signal, it is by no means obvious how far from equilibrium this noise can move the system. Fitting a curve, extending to infinity, through a few observed states does not in any way indicate how far the corresponding energy situation makes extrapolation along this curve physically meaningful. Maybe the system could not produce an event labelled a 1,000-year flood at all, maybe it would be capable of generating much higher floods with a much higher probability. This cannot be decided by fitting one or other mathematical function to a given set of points but only on the basis of a good understanding of the behaviour of the physical processes involved. There may well be two similar sets of annual maxima, with similar best-fit curves, but coming from different climatic regions, for instance one dominated by snow melt floods and the other caused by convective storms. Is there any reason to assume that a 1,000 or 10,000-year flood will be of a similar magnitude in both of them?

Another point which is extremely important is a possibility that long before the processes reflected in a short historic record could, through their random interactions, produce a really extreme flood, another process, perhaps not at all active during the period of record, could intervene and produce a much bigger flood. A case in point may be eastern Canada where most annual maxima are generated by snowmelt but occasionally a hurricane can stray into the region and cause a flood much larger than is conceivable via the snowmelt mechanism.

It is becoming increasingly more acceptable to extrapolate the fitted 'probability distributions' of floods beyond limits considered reasonable in the past. While one can hardly find a reference to anything larger than a 100-year flood in the literature from the turn of this century, a 1,000-year flood was quite a common concept 30–40 years ago and today a 100,000-year or a million-year flood may seriously be invoked in connection with flood protection of toxic-waste dumpsites, strategic military installations, or nuclear power plants. While it is easy enough to extrapolate the 'best fit' curves to these extremes, it is also easy to arrive at complete absurdities in this manner. For example, one could 'estimate' a million-year flood with peak flow of x m^3/s on the Manitou River on the Manitoulin Island in Lake Huron. But it may well be that under conditions necessary to produce such a flood, the whole Manitou River, and even the Manitoulin Island, would disappear so that no such peak flow could ever arise there.

Without an analysis of the physical causes of recorded floods, and of the whole geophysical, biophysical and anthropogenic context which circumscribes the potential for flood formation, results of flood frequency analysis as described above, rather than providing information useful for coping with the flood hazard, themselves represent an additional hazard that can contribute to damages caused by floods. This danger is very real since decisions made on the basis of wrong numbers presented as good estimates of flood probabilities will generally be worse than decisions made with an awareness of an impossibility to make a good estimate and with the aid of merely qualitative information on the general flooding potential. This warning is more relevant now than it may have been in the past because of the increasingly apparent nonstationarity of the climatic signal which suggests that the assumption of stationarity is no more defensible even on a very short time scale of two or three decades. Other aspects that should be considered in assessing the potential of extreme floods are the land-use changes in the basin, effects of the fluctuation of ocean heat storage on local weather patterns, effects of water storage in the basin on the flood response to precipitation, effects of volcanic eruptions on flood climates, effects of crustal movements on watershed boundaries (especially in flat regions such as, for example, the Great Lakes Basin), changes in drainage areas for small and large floods (e.g. the case of the Santa Anna River basin in

California), effects of the changing patterns of atmospheric deposition on the melting of snow and glaciers, effects of tectonic activity on the safety of large dams, effects of various periodic signals (e.g. Currie and O'Brien, 1988) and perhaps other phenomena (e.g. see Geophysics Study Committee, 1978). A more detailed criticism of current practices was given in Klemeš (1987a), while some promising new physically based approaches to assessing the flood potential can be found in Baker *et al.* (1988). Unfortunately, there has not been much innovative statistical thinking about floods besides the rudiments proposed by Todorovic and Zelenhasic (1970), and Eagleson (1972).

Extreme Droughts

The prevailing attitude to the probabilities of extreme droughts is essentially the same as to those of extreme floods and is well summarized in the following statement: 'Just as the Gumbel and the Frechet distributions are a natural choice for flood data because they are tailored to the properties of sample maxima, the Weibull distribution is an attractive choice for data which can be interpreted as minima of samples' (Beran and Rodier, 1985, p.75). Thus the only tangible difference is the type of the mathematical function of the likely 'best fit' from which the 'probabilities' are estimated. At most, the higher persistence of drought-related data is sometimes taken into account to reduce the return periods compared to those corresponding to random samples (which is how flood samples are treated).

If the current practice of estimating probabilities of extreme floods by extrapolating the upper tails of 'best-fit distributions' is bad, then estimating probabilities of droughts by extrapolating their lower tails is simply awful. While floods do exhibit some features of randomness in that they appear suddenly, may affect only small areas, and their long-term patterns indicate a high level of noise, droughts appear to represent a markedly different type of process. Typically, they develop slowly over time (they have been labelled a 'creeping' phemomenon), last much longer than floods, affect large and often enormous territories, tend to show cyclic patterns, etc. For example, the 1930s drought affected three continents and extended from north-east America through Europe to west Siberia (Shiklomanov and Markova, 1987); the current Sahel drought has lasted for about 20 years; during the past 80 years South Africa experienced four drought cycles centred around the 1910s, 1930s, 1950s and 1970s (Tyson and Dyer, 1978), etc.

From a physical point of view, droughts usually represent a process of a different order than do floods and 'symmetry' between minima and maxima of the same process generally cannot be invoked. The general relationship between the processes of floods and droughts can perhaps be best illustrated with the following simplified example. If floods were represented by the maxima of a normalized function of time, $x(t)$, then droughts would be represented much better by the minima of a normalized function $y(t) = \int x(\tau)d\tau$ than by the minima of $x(t)$ itself. Depending on the specific type of drought 'meteorological, hydrological, agricultural, socio-economic, etc.), its meaning will more or less diverge from that of min y because droughts arise through accumulation of water deficits in 'hydrologic reservoirs' of different scales and governed by different 'operating mechanisms' (Klemeš, 1987b). Thus the difference is not merely between maxima and minima of a given sample, but between their probabilistic meanings. Even if two samples of historic floods and droughts could be fitted with the same distribution functions, the same relative frequencies on the two curves would be associated with different probabilities. This is not merely a matter of a smaller or greater persistence, but also of phase shifts between the two processes, their propensity to quasi-periodicity, etc. Such differences would be even more pronounced in 'socio-economic' droughts which may be of the type of min z where the (normalized) $z(t) = \int y(\tau)d\tau$. The structure of the relevant processes may become quite complex (see Klemeš and Klemeš, 1988, for details) and their probabilistic interpretation even more so. It therefore may be dangerous to rely on drought 'probabilities' obtained by 'frequency analysis' as it is routinely practiced.

Conclusions

To increase our understanding of probabilities of extreme floods and droughts and to improve their estimates, more would be gained by the study of geophysical and anthropogenic processes than by the present preoccupation with the subtleties of distribution fitting which merely diverts talent from the task of shedding light on the dark areas to a futile search for the key to nature's secrets under the proverbial street lamp. As one mathematician aptly put it in a more general context: 'We may write down equations, and nature may – at some level – obey them, but nature is not obliged to restrict herself to those solutions that our overgrown monkey intellects can write down explicitly. And so mathematics must pay attention to what really happens, rather than assume that nature conspires to make human calculations easy,' (Steward, 1988).

REFERENCES

BAKER, V.R., KOCHEL, R.C. and PATTON, P.C. (Eds) (1988) *Flood geomorphology*. John Wiley & Sons, New York

BERAN, M.A. and RODIER, J.A. (1985) Hydrological aspects of drought. Studies and reports in hydrology, No. 39. Unesco/WMO. Paris/Geneva

CURRIE, R.G. and O'BRIEN, D.P. (1988) Periodic 18.6-year and cyclic 10 to 11 year signals in northeastern United States precipitation data. *J Climatol* 8: 255–281

EAGLESON, P.S. (1972) Dynamics of flood frequency. *Water Resour Res* **8:** 878–897.

GEOPHYSICS STUDY COMMITTEE (1978) *Geophysical predictions.* Nat Acad Sci (USA) Washington DC

KLEMEŠ, V. (1978) Physically based stochastic hydrologic analysis. *Adv Hydrosci* **11:** 285–355

KLEMEŠ, V. (1986) Dilettantism in hydrology: transition or destiny? *Water Resour Res* **22:** 177S–188S.

KLEMEŠ, V. (1987a) Hydrological relevance of flood frequency analysis. In: *Hydrological frequency modeling.* Singh, V.P. (Ed.) D. Reidel, Dordrecht, pp. 1–18

KLEMEŠ, V. (1987b) Drought prediction: a hydrological perspective. In: *Planning for drought* Wilhite, D.A. and Easterling, W.E. (Eds) Westview Press, Boulder, pp. 81–94

KLEMEŠ, V. and KLEMEŠ, I (1988) Cycles in finite samples and cumulative processes of higher orders. *Water Resour Res* **24:** 93–104.

SHIKLOMANOV, I.A. and MARKOVA, O.L. (1987) A world survey of problems of water availability and transfer (in Russian). Gidrometeoizdat, Leningrad

STEWART, I. (1988) Nature and the monkey intellect (Review of 'Mathematics and the Unexpected' by I. Ekeland). *New Scientist* **119:** 59

TODOROVIC, P. and ZELENHASIC, E. (1970) A stochastic model for flood analysis. *Water Resour Res* **6:** 1641–1648

TYSON, P.D. and DYER, T.G.J. (1978) The predicted above-normal rainfall of the seventies and the likelihood of droughts in the eighties in South Africa. *S Afr J Sci* **74:** 372–377

Control Training of
Water Resources Systems under
Simulated Flood Conditions
—

VLADIMÍR BLAŽEK[1] and JAN KUBÁT[2]
[1]*Hydroprojekt Praha,*
[2]*Czech Hydrometeorological Institute, Prague, Czechoslovakia*

Introduction

EXTREME flood situations that cause considerable damage occur only rarely. In spite of that, it is necessary to be perfectly prepared for dealing with precisely these situations. The problem consists of the fact that the decision-makers (staff of water authorities and of the control centres of water works operators) have very little personal experience in this respect, due to the low periodicity of these situations. They make decisions in situations most of them have never been through before, often pressed for time and by the higher-level bodies. At the same time, their decisions may have far-reaching consequences.

In an effort to prepare the personnel of decision-making bodies for the execution of their duties under extreme flood conditions, methods of training in simulated situations are being developed in Czechoslovakia. The underlying philosophy is analogous to that of simulation games in other fields, for example economy. A computer simulates the course of a flood situation, outputting the essential data which the person at a control centre receives in the real time mode. The trainee takes action through his decisions and is evaluated by the results he achieves at control sites. The entire training scheme may be devised in the simultaneous form when several decision-making centres compete under identical input conditions. It may have the interactive form when co-operation of several decision-makers is simulated, such as the hydrological forecasting centre, the control centre, and the flood authority.

Several simulation games have been tested in Czechslovakia, with teaching, training as well as research purposes in mind. Our experience has shown that the method is a highly effective one, with good prospects for further development.

Flood Protection

Flood protection is one of the most important tasks of the water management sector in Czechoslovakia. Due to the uneven hydrological regime of water streams, floods occur relatively frequently, especially in the spring and summer months. Average annual damage caused by floods is estimated at 500 million Cz.crowns, however, cases are not rare when damage resulting from a single large flood exceeds 1,000 million Cz.crowns.

Flood protection measures include regulating and increasing the capacity of water courses, protective dykes, and reservoirs which can actively help to influence the form of flood waves. Apart from single-purpose retention reservoirs, most of the multi-purpose reservoirs are employed in flood protection. Although only about five per cent of the length of water courses is under the direct influence of reservoirs, most of these stretches are located in economically and socially valuable areas, for example the capital of Prague, the spa resort of Karlovy Vary, industrial areas in northern Bohemia, and farming regions in southern Moravia and Slovakia.

Flood Control Service

When a region is affected by floods, the system of flood control service is activated and operates according to previously drafted flood plans. The flood control service involves operators of courses and water works, the hydrological forecasting service, and also managers of endangered facilities in the inundation area. If need be, other authorities participate. Under the Czechoslovak Water Act, the responsibility for managing the flood control service goes to the water authorities of National Committees, the terms of reference of which also include the right to impose extraordinary measures.

In Czechoslovakia, specialized organizations (the River Boards) are in charge of major water courses and most dams. Attached to these river boards are control centres for the operative control of water courses and structures located on them. To illustrate, one control centre is responsible, on average, for 15,000 km^2 of area, 3,500 km of rivers and several dozen dams and gated weirs. Some control centres have their own automatic data collection systems that cover their region, while the building of automatic measuring networks for other centres is being prepared. The River Board control centres' tasks are especially important in extreme situations such as floods, droughts, water pollution accidents and technological emergencies.

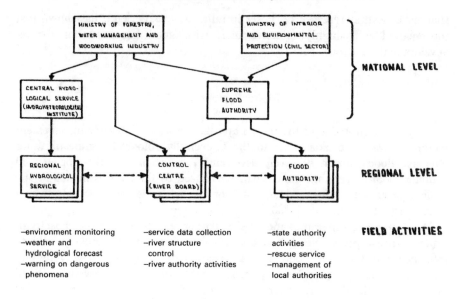

FIGURE 1 *Scheme of the flood control service*

Training on Model Situations

The successful control of flood situations requires careful material and organizational arrangements by the flood control service. There is a need for field data collection, preparing flood plans and operational rules for water works and water systems, evacuation and rescue plans, watch duty schedules, etc. What is equally important, is the readiness of people, especially those assigned to make decisions about the measures to be carried out within the flood control system. In this respect, the levels of the control personnel's qualifications are not fully adequate to the demands placed on them during a flood situation – i.e. to decide about complex technico-organizational questions within short periods of time, frequently when lacking all the information needed and, sometimes, under stress and the pressure from state and political authorities. Extreme situations with low periodicity are usually beyond the involved personal experiences of people. However, the decisions made by those persons may have serious economic, social and other consequences.

One of the useful methods of preparing people for contingencies seems to be training in control activities on model situations. Such models may concern flood situations, extreme drought, accidents on water works, water pollution accidents, extreme ice phenomena on water courses and the like. Situations that have never happened before may be simulated advantageously, thereby putting to the test people's behaviour as well as that of water structures systems under extreme conditions. In Czechoslovakia, a general methodology has been worked out, and applied so far to several flood simulations of the Ohře and Bílina river systems.

Flood Conditions Simulation

The training scheme is based on simulating a flood situation for one or more teams included in the flood control service. The teams are to carry out their duties in the same way as under actual flood conditions. In this country, three main partners co-operate during a flood:

- operative control centre of the manager of the water courses in question (the River Board control centre);
- the relevant workplace of the hydrological forecasting service (Hydro-meteorological Institute);
- local water authority (National Committee).

The training entails the simulation of an information system that supplies the personnel with the required meteorological, hydrological and operational data on the state of the river system on which the training is being realized. The purpose is to create for the personnel, if possible, real-life working conditions, including the option of using real working aids and tools as in actual operation. Individual teams carry out activities within their competence and pre-defined rules, i.e. they issue hydrological prognoses, decide about the actions to be taken on dams, order evacuation or other steps to prevent flood damage. The water system's response to their decisions is modelled in the next simulation step.

Types of Simulation (Training)

Simple simulation

Only one decision-making team linking to the water resources system is simulated. As a rule, it is the control centre and it decides about actions to be taken in dams. The system's behaviour is simulated on a personal computer which facilitates simple and flexible contact between the trainees and the system. This type of simulation makes it possible to test different alternative decisions and their consequences; analyse the system's behaviour in made-up meteorological and hydrological situations; or verify retrospectively an actual situation from the past.

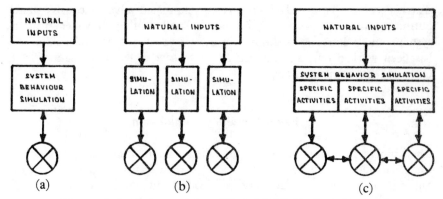

FIGURE 2 *Types of simulation training; (a) simple simulation, (b) simultaneous simulation, (c) interactive simulation.*

Simultaneous simulation

The links of several decision-making teams of one type to the water resources system are simulated in parallel. All participants deal with the same situation under the same conditions, and the success rate of their decisions can be evaluated on the basis of, say, decrease of flood discharges at control sites, magnitude of flood damage or penalty points. Since the participants compete with each other, the training becomes an interesting game for them. This type of simulation is suitable above all for teaching purposes. It is more demanding on organization because multiple simulations of the system's responses are required. An advantageous configuration is one central computer with terminals for each team.

Interactive simulation

This entails links of several co-operating decision-making teams to the water resources system as well as their interrelations, for example, hydrological service – control centre – water authority. It is the most complex type of training, and the one which best approximates training conditions to the real-life working conditions. It makes it possible to test relations between the individual members of the flood control service and train their collaboration. However, arrangements for such training are complicated because, in general, not all links can be simulated on the computer. The trainees' activities are evaluated by a referee who must be an experienced member of staff of the flood control service.

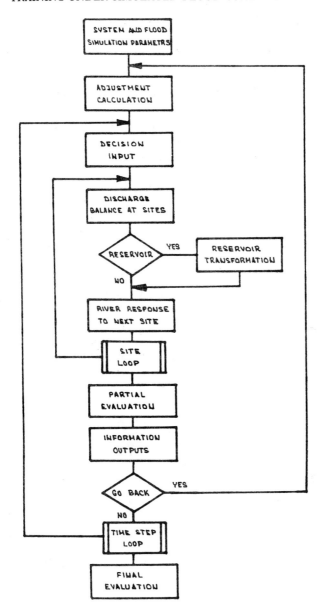

FIGURE 3 *Scheme of a simple simulation model*

How a Training Scheme is Prepared

Each training course requires careful preparation by a team of experts. The preparation comprises several stages:

– define objectives of training and select type of simulation;
– delineate the water resources system;
– analyse the information system, determine the variables to be monitored;
– arrange the meteorological and hydrological situation;
– supply complementary information and inputs;
– identify and verify the computer model;
– define criteria for evaluation of participants;
– test-run the simulation.

An important aspect is to estimate correctly the duration of training. Assuming that the overall duration of the simulation itself should not exceed eight hours, the ratio of simulation and real time is 1:10 to 1:20 for a flood situation that lasts several days. In one simulation step, the trainees should be allowed about 15 minutes for making a decision, and up to 30 minutes for deciding about critical issues. A reasonable length of a run is 15 to 20 steps.

Prior to the simulation itself, the participants must receive introductory instruction, and be acquainted with the river system, control regulations and evaluation rules. They must have the possibility of studying basic materials and/or conducting preparation on their own. When the simulation run comes to an end, there should be time enough to evaluate and discuss it. It is useful to include in the training scheme specialized lectures on relevant topics. Optimally, the training program would be scheduled for three days or more, if necessary.

In Czechoslovakia, training on model flood situations was first tested in 1982 in the form of interactive simulation by three teams in co-operation. It was repeated in 1984 on a different river system under different conditions. In 1986, a computer model was developed for simultaneous simulation, which was also repeatedly used for teaching purposes. In 1986 and 1988, training was organized for participants in a UNESCO international postgraduate hydrological course. A simple model implemented on a personal computer with graphics outputs has been derived from the simultaneous simulation computer model.

Example of an Actual Training Course

An illustrative example is the simulation of a summer flood situation in the Ohře river basin. Its river system covers 5,600 km^2. For training purposes, it was simplified to include only four major rivers. The trainees decided about the handling of four reservoirs having defined protective zones. A fixed simulation time step of six hours was used. In each time step, the trainees received information on water levels in all four reservoirs, discharges at 11 control sites and precipitations at 12 gauging sites. Every 12 hours, they received a simulated weather forecast supplemented with a synoptic chart and/or met-radar pictures. The success rate of the flood control was evaluated on the basis of discharges at three control sites downstream. Depending on the maximum discharge achieved, the participants scored penalty points on the basis of a loss function which expressed the extent of flood damages in a given locality. Other penalty points may have been scored for exceeding the maximum limit of water levels in the reservoirs that could have been critical to the dam's safety.

For this simulation, a summer flood was selected that might affect the whole basin with a relatively realistic probability (10 to 20-years flood). For heavier floods, the reservoirs' capacities are not sufficiently large to effectively influence the flood wave at control sites. A number of participants, beginners as well as experienced specialists, dealt with the situation from the position of control personnel. The success rate of control amounted to 20 to 80% of that of optimum control derived via theoretical analysis.

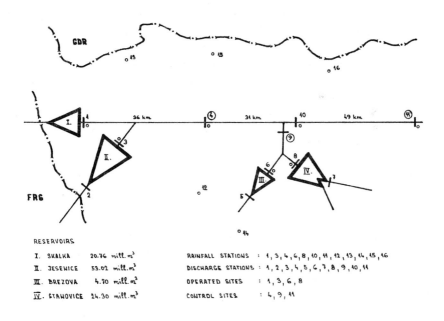

FIGURE 4 *Water resources system of the upper Ohře river*

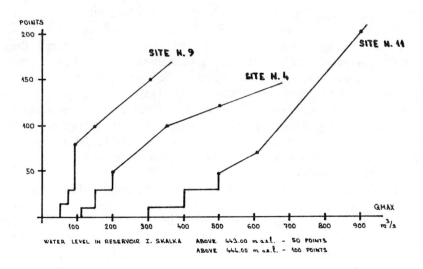

FIGURE 5 *Evaluation at control sites (penalty points)*

Further Evolution of the Method

Currently, a model is being prepared for interactive simulation of a flood situation. Again, three teams are being considered: hydrological service – control centre – local flood authority. In the next stage, a fourth team is to be added – meteorological service. Each of the teams should have a personal computer or a terminal at its disposal. The basic model, which simulates natural inputs and the system's behaviour, will be run in parallel on all computers. Apart from this, each team will have access to auxiliary subroutines for its own specific performance. The individual teams' activities will affect one another, particularly in the meteorology-hydrology river-control-flood service direction. The final criterion of success will be the magnitude of flood damage and that of the flood-service's financial outlays. The simulation is being prepared for the Labe-Vltava river system which covers the territory of Bohemia. This river system will be simplified to include eight major rivers, five principal reservoirs and 20 discharge gauging sites. The training scheme is expected to be implemented in 1990.

Judging by the experience gained so far, training on model situations appears to be an effective method of improving the qualifications of personnel. It is an interesting approach and helps to keep them active. However, preparations require a considerable amount of effort, and so far have depended on the enthusiasm of a few individuals. To introduce a regular regime of training of broader scope will require systematic development of the method at

VLADIMÍR BLAŽEK AND JAN KUBÁT

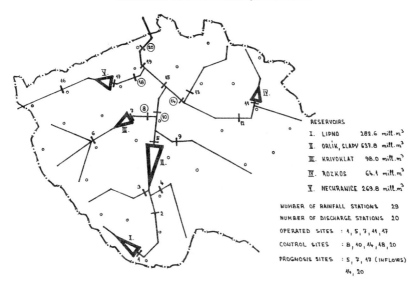

FIGURE 6 *Water resources system of the Labe-Vltava rivers*

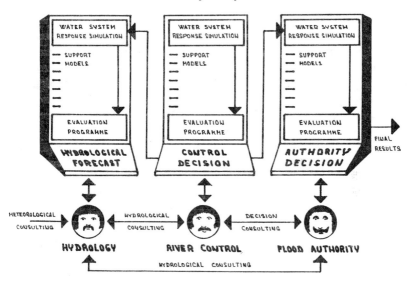

FIGURE 7 *Interactive simulation on three personal computers*

a specialized workplace with professional backing. Apart from teaching and training purposes, the method is useful in the verification of techniques and regulations pertaining to the control of water resources systems, the effectiveness of the information system used, and the mutual relations between the individual persons involved in the control. It may also be a contribution to the further development of the theory of water resources system control.

Hydrological Aspects of a Flood Management Expert System

P. BAKONYI and P. BARTHA

Research Centre for Water Resources Development,
VITUKI, Budapest, Hungary

Introduction

FLOOD fighting has a long tradition in Hungary. In the 19th century a large part of its territory was inundated once or twice a year. The big anti-inundation works started in the first half of the last century. Up to now a 4,186 km long dyke has been built along the main rivers and their tributaries. The protected area is 21,200 km^2 – one fifth of the territory of Hungary. During the construction of dykes, extensive river training helped the carrying away of flood waves.

Although construction of flood prevention establishments has almost been finished, flood management has still not lost its importance. Flood waves have become higher and longer – due to human intervention within the catchment area – and may overtop and drench the levees. The levees have to resist excessive water pressure, sometimes for weeks. With the passing of time, the resistance of the levees weakens, which may result in levee breaches. The danger to human life and the value of the protected area make efficient flood management essential here.

Flood fighting activity in Hungary is co-ordinated by the Ministry of Environment and Water Management. Operational flood protection is done at the level of the 12 District Boards of Environment and Water Management. The head of each board is responsible for flood prevention. To help this activity a so-called Flood Management Expert System (FLOMEX) was built. It can collect, store and retrieve vital information about the catchment area and the situation during the flood fighting. Its expert capability helps the head of flood protection to verify the consequences of his decisions.

In this paper a short presentation of FLOMEX will be given. As flood prevention is directly related to hydrological conditions within the catchment area more emphasis will be put on the presentation of flood forecasting and flood routing modules of FLOMEX.

Flood Management Expert System

The Flood Management Expert System is aimed to assist in operational flood protection. It is an integrated software package written for IBM PC AT/XT. The basic idea behind FLOMEX is to collect information from the watershed, evaluate the situation and, using the reasoning capability of the expert system, recommend appropriate action. To fulfil this intention FLOMEX performs the following tasks:

- data collection (precipitation, water level elevation, condition of dykes, number of people in action, etc.);
- flood forecasting and backwater calculations;
- analysis of the situation (overtoppings, rising of water levels, weakening of dykes, etc.);
- suggestion of action to be taken (increase the warning level, evacuate an area, etc.);
- acknowledge the decision of the head of the protection and warn him about its consequences.

The operation of FLOMEX is based on data collection. Hydrological and other information is transferred to the protection headquarters through the normal ways (telephone, telex, automatic data acquisition devices, etc.). Then information is fed into the computer – only manually at present – and stored on its fixed disk. The data processing is done partly parallel to the loading of information and partly after it.

Hydrological (and other) data are checked for typing errors and range errors. Only error-free data are stored in the database. Parallel to the storing of data they are checked from the point of view of flood prevention. Rising water levels and high precipitation are noted and a warning is given to the user. Most of the data are only used during the consultation with FLOMEX as one of its major tasks is to make flood forecasting.

The flood forecasting module of FLOMEX transforms the precipitation data into runoff. The model provides direct data for flood management – water levels at the gauge stations – and input data to the backwater calculation module. This later provides the head of flood protection with longitudinal profiles of water level elevations to make safety checks possible between gauge stations, too. FLOMEX has some more very interesting features but in accordance with the aims of the Technical Conference on the Hydrology of Disasters, only these two main modules will be presented below.

Hydrological Module of the Flood Management Expert System

The rainfall-runoff model was developed in the framework of the Project of the Development of Hydrological Forecasts for Hungary to serve in the system of operational forecasting on Hungarian rivers. The basic idea behind the model is to divide the rainfall-runoff process into two subprocesses:

- the formation of effective rainfall, i.e. the difference of actual rainfall and interception plus evaporation losses, which initiates surface and subsurface runoff and infiltration into lower zones;
- the routing of surface and subsurface runoff through river bed network and the subsurface layer of soil into the downstream section of the basin.

The GAPI model consists of three main modules:

- *Module 1* esimates the effective rainfall and separates surface, subsurface and underground flow (first subprocess);
- *Module 2* routes the surface, subsurface and underground flow to the reference station taking into account the effect of storage capacities;
- Module 2 describes the stochastic processes which are not modelled by in the deterministic modules.

In Module 1 the effective rainfall is expressed through the volumetric runoff coefficient:

$$\alpha_i = \frac{\sum\limits_{j=i-M+1}^{i} Q_j - MQ_0}{\sum\limits_{j=i-M-T+1}^{i=T} u_j}$$

where

α_i volumetric runoff coefficient,
Q_0 base flow rate ($m^3\ s^{-1}$),
T time lag of runoff to rainfall (days),
M period of rainfall and runoff integration (days),
Q_i flow rate at the downstream station ($m^3\ s^{-1}$), and
u_j rate of precipitation ($m^3\ s^{-1}$).

The rate of effective rainfall largely depends on seasonal changes of catchment humidity conditions, i.e. indirectly on characteristics of antecedent rainfall. It is evident that the earlier a rainfall event took place, the less it affects the current conditions of the catchment. To describe this kind of retention the antecedent precipitation index (*API*) is widely used in hydrology. This is, in fact, a series of weighting coefficients that diminish backwards in time. The model GAPI applies the following exponential form of the precipitation index:

$$APUI_i = \sum_{l=0}^{n} u_{i-l}\, e^{-la}$$

The *API* is calcuated on the basis of effective daily rainfall values.

The ratio of surface runoff is expressed through the probability of *API*. This approach is applied owing to the probabilistic character of the relationship though it has also a strong deterministic basis.

The distribution of *API* was investigated for more than 10 river basins and the following expression gave the most suitable form for the density function:

$$f_P(API) = \frac{\lambda^P\, API^{P-1}}{\Gamma(P)} \cdot e^{-\lambda API}$$

This distribution approaches exponential distribution for $P = 1$ and Gaussian distribution for $P > 50$.

Assuming that there is a dependence of the effective catchment area on the probability of *API*, the ratio of surface runoff can be estimated if the probability of *API* is known:

$$A_{f,\,i} = f\,[P(API)]$$

and the interflow ratio is:

$$A_{\alpha,\,i} = 1 - A_{f,\,i}$$

Finally, Module 1 separates the total volume of runoff into three parts:

– the surface flow

$$u_{1,\,t} = u_t\, A_{f,\,t}\, \alpha_t$$

– the subsurface flow (or the time variant part of it)

$$u_{2,\,t} = u_t\, A_{\alpha,\,t}\, \alpha_t$$

– the base flow (or the time invariant part of the underground inflow)

$$u_{3,\,t} = Q_0 = \text{const.} = Q_{min}$$

The surface and subsurface flow are routed to the downstream station by Module 2 (the base flow is taken as constant for the whole hydrological year). The Discrete Linear Cascade Model (DLCM) – Szöllösi-Nagy (1982) – was implemented for the routing of the surface and subsurface flow. For the sake of simplicity the same model (with different parameters) serves for the routing of the two flow components.

The structure of Module 2 is shown in Figure 1 and consists of two parallel cascades. The first cascade with parameters (n_1, K_1) routes the surface flow and the second one with parameters (n_2, K_2) the subsurface flow. The DLCM can be expressed in the following form:

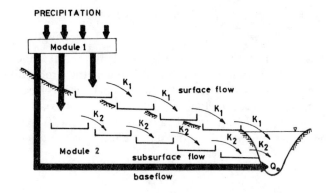

FIGURE 1 *Modular structure of the GAPI model*

$$x_{i,\,t} = \Phi_i(\Delta t)\, x_{i,\,t-\Delta t} + \Gamma(\Delta t)\, u_{i,\,t}$$

$$y_{i,\,t} = H_i\, x_{i,\,t}$$

with $i = 1$ for surface and $i = 2$ for subsurface flow. The state transition (Φ_i) and the input transition (Γ_i) and the output (H_i) vectors are described in Szöllösi-Nagy (1982).

The model GAPI produces the discharge rate at the downstream station for each discrete time step as a sum of the routed values of surface, subsurface and base flow:

$$\hat{y}_t = y_{1,\,t} + y_{2,\,t} + Q_0$$

The stochastic submodel of GAPI (i.e. Module 3) is of simple continuous correction type known from the practice of hydrological forecasting. The procedure is based on the differences of calculated 'forecast' values and the measured values for each discrete time step.

$$\varepsilon_t = \hat{y}_t - y_t$$

The correction for the lead time of the forecast:

$$\Delta y_{t\,+\,\Delta t} = -\,\varepsilon_t$$

and the corrected forecast:

$$\hat{y}_t^{(k)}{}_{+\,\Delta t} = \hat{y}_{t\,+\,\Delta t} + \Delta y_{t\,+\,\Delta t}$$

This continuous correction serves for the interests of error handling and gives satisfactory results even in the case when in the error sequence only the one-step autocorrelation is significant (Szöllösi-Nagy *et al.*, 1983).

Hydraulic Module of the Flood Management Expert System

The backwater curve calculation in the Flood Management Expert System is based on the solution of the complete de Saint-Venant equations:

$$\frac{\partial Q}{\partial x} + \frac{\partial A}{\partial t} = 0$$

$$\frac{\partial z}{\partial x} - \frac{\alpha' Q^2}{gA^3} \frac{\partial A}{\partial x} - \frac{\alpha'' Q}{gA^2} \frac{\partial A}{\partial t} + \frac{\alpha' Q}{gA^2} \frac{\partial Q}{\partial x} + \frac{\alpha''}{gA} \frac{\partial Q}{\partial t} + \frac{Q^2}{K^2} = 0$$

where:

Q discharge $(m^3\ s^{-1})$,
z water level elevation (m),
x longitudinal co-ordinate (m),
t time (s),
A wetted area (m^2),
K conveyance coefficient $(m^3\ s^{-1})$,
g acceleration due to gravity $(m\ s^{-2})$, and
α', α'' dispersion coefficients.

The solution used is the well known Abbott-Ionescu implicit finite difference scheme (Cunge, Holly, Verwey, 1980). The discretization of the set of differential equation is done on a so-called staggered grid (i.e. calculating only one unknown – discharge or water level – at each grid point). This type of discretization results in a very effective and fast solution procedure. The resulting non-linear system of equation is solved by the Netwon-Raphson iteration.

As Hungary is basically a flat country the backwater effects of the tributaries cannot be neglected. Therefore the solution was extended to 'tree-like' river systems (i.e. no loops). For this purpose special continuity conditions had to used at the junctions of rivers.

The most difficult problem in modelling flood waves is the prescription of boundary conditions. For the model applied in the FLOMEX as many boundary conditions are needed as there are open ends of rivers (i.e. two for a single river, three for a river with a tributary, etc.). The upstream boundary conditions are discharge time series defined by the GAPI rainfall–runoff model. The downstream boundary condition – considering the Hungarian hydrological conditions – is a rating curve. This is almost a non-reflecting boundary which lets the water flow out of the modelled river reach without significantly effecting the water levels.

The solution of the Saint-Venant equations requires knowledge of the initial conditions. In FLOMEX there are two ways of defining them. At the start, the initial conditions are calculated as a steady-state, gradually varying backwater curve. Once the program is run it writes the water levels and discharges into a file that can serve as the initial conditions for the next calculation. This technique is called 'hot-start'.

During the daily operation of FLOMEX the hydrological and hydraulic models are only activated once a day. Having done the forecasting the backwater curves are calculated. At each stage the calculated water levels are compared to the warning levels at the gauge stations and to the level of levees. In case of the water levels exceeding warning levels or the levels of levees FLOMEX gives a warning to the user and the expert module advises the necessary actions to be taken.

In the simulation phase the models are activated according to user needs.

Future Plans

The development of FLOMEX has not yet been finished. To improve its hydrological applicability, two more features will be added:

– levee breaching model coupled to the solution of the Saint-Venant equation to model the effects of the use of emergency flood reservoirs,
– incorporation of meteorological radar data into the rainfall-runoff model.

With these improvements the lead time of the forecast can be increased by a few hours and the consequences of an artificial or natural levee breach can be simulated. Both will provide significant support to the activities of the head of protection.

REFERENCES

BARTHA, P. and HARKÁNYI, K. (1986) Rainfall-runoff forecasting model 13th Conference of the Danubian Countries on Hydrological Forecasts, Belgrade, September 1986.

CUNGE, J.A., HOLLY, F.M. JR. and VERWEY, A. (1980) *Practical aspects of computational river hydraulics.* Pitman Advanced Publishing Program, Boston

SZÖLLÖSI-NAGY, A. (1982) The discretization of the continuous lineal cascade by means of state space analysis. *J Hydrol* 58: 223–236.

SZÖLLÖSI-NAGY, A., BARTHA, P., and HARKÁNYI, K. (1983) Microcomputer based operational hydrological forecasting system for river Danube. Technical Conference on mitigation of natural hazards through real time data collection systems and hydrological forecasting. Sacramento.

The French Policy for Prevention of Flood Damage

P.A. ROCHE

Water Planning Authority,
Secretariat of State for the Environment,
Directorate for Water and the Prevention of Pollution and Risks,
Neuilly-sur-Seine Cédex, France

THE CHOICE between preventing natural disasters and protecting against them would at first sight appear simple. However, for a number of years now the State's anti-flood policy has taken the form of a judicious alchemy of time-honoured remedies and remarkable innovations in both the technical and administrative/legal fields.

Floods are the most serious of all the world's natural disasters. Floods cause much more damage than tropical cyclones or earthquakes, and lead to the loss of thousands of human lives every year. Localized flooding, generally arising from violent storms, can wreak havoc in urban areas, as was seen in Nepal in 1987 (400 dead) and in Rio de Janeiro in 1988. The probability of such occurrences must be taken as a crucial factor for town planning considerations.

The risk of widespread flooding over very large basins, which has been the cause of major disasters (100,000 died in the Yangtze flood in China in 1911), constitutes a limiting factor to national development, especially as regards town planning. Such natural restrictions may be partially overcome by protective measures but cannot be totally eliminated. Countries in which vast natural flood plains have been reclaimed by dyking must live under the constant threat of dyke failure. This is the case in Hungary, a quarter of whose territory is made up of former fllood plains which now house virtually all the country's major settlements and industry.

In France, such phenomena are less frequent but may nevertheless have tragic consequences: 23 people died in the Grand-Bornand disaster of July 1987 (see Appendix 1). More recently, Nîmes was flooded in October 1988; and the floods of the winter of 1982/83 cost nearly 2 million francs in direct damage.

In view of the importance of the problem at hand and the demands of economic development, the Secretariat of State for the Environment and the Secretariat of State for the Prevention of Pollution and Risks, co-ordinators of State undertakings in the field of Water and Risk Prevention, pursue a compromise policy tailored to the particular needs of the French nation. This combines a risk prevention policy aimed at reducing the vulnerability of people and property, and a protection policy applied to those areas in which land use is considered essential for economic reasons.

Prevention of Flood Damage

The French policy for flood damage prevention has the following goals:

- To provide prompt and precise information as to flood risks;
- To improve and perfect town-planning in flood-prone areas
- To control flood phenomena by means of public works constructions wherever this is technically feasible and economically justifiable.

Flood Warning

Improved public-service awareness

If the population is kept informed of rising water levels, it will be better able to take the appropriate measures to protect lives and property. For this reason, although under no legal obligation to do so, the State runs a flood forecast and warning service covering 16,000 km of rivers.

The first flood-warning service came into being in 1854 in the Seine basin. Such services were later adopted wherever justified by socio-economic considerations, provided that they could prove sufficiently reliable; on watercourses with rapid response times (of the order of hours, say), flood warnings would be of dubious utility as it would be unlikely that riverside dwellers could be warned early enough. At the present time, France has 53 Flood Warning Services (run, at *Département* level, by Public Works Directorates, Forestry and Agriculture Directorates and Navigation Boards).

The public alert system was recently reformed by the ministerial decree of 27 February 1984. The methods used for giving the alarm are now set out in local regulations established by the *Préfet* of each *Département*, who is responsible for alerting the mayors of the towns and villages concerned. Once alerted, the mayors have access via a confidential number to a telephone answering machine (operating in transmit/disseminate mode) which issues messages regularly updated from the *Préfecture*.

High-tech data transmission systems

To accompany the administrative reform, it was decided to go ahead with an extensive modernization program aimed at removing the need for employing human observers to monitor river levels and transmit this data to the flood warning centres.

Besides the difficulties involved in recruiting observers who are expected to offer permanent availability (certain watercourses may rise one metre in an hour and require constant surveillance), the use of human observers also entails problems of accuracy and reliability. For this reason, automatic monitoring stations have been developed; these will measure and store hydrometeorological data at the riverside site itself and then transmit this data by radio, satellite or telephone line to the Flood Warning Services where it is needed for forecasting purposes. If water level or rainfall thresholds are exceeded, the central site computer will warn the duty officer, who will in turn inform the *Préfet* with a recommendation to raise the alarm.

The modernization program will cost about 160 million francs over eight to 10 years. This cost will be equally divided between the State and local government bodies (*Régions, Départements, Communes*) As yet, 450 out of the projected total of 850 monitoring stations have been automated. First to be modernized were the largest basins, i.e. those of greatest impact (Garonne, Loire, Seine). The next, smaller, networks to be modernized include those of Hérault, Finistère, Saône, Moselle. These will all be provided with the same equipment, defined as the result of nation-wide consultation. The program is to reach completion in 1991-1992 (see Figure 1 and Appendix 2).

Emergency measures

In each *Département*, the implementation and co-ordination of emergency measures for dealing with floods will come under either the responsibility of the mayor or that of the *Préfet* of the *Département*, depending on the gravity of the situation at hand. Under the terms of his general mandate, the mayor is empowered to take all measures necessary and muster such assistance as may be required to prevent or cease floods and any other calamities. In the event of need or catastrophe of exceptional magnitude, the *Préfet* takes control of emergency operations, with powers to use all public or private means as might be available. All *Département*-scale emergency plans, including those for coping with floods, come under the responsibility of the *Préfet*.

Compensation for flood victims

Since the law of 13 July 1982, concerning compensation for the victims of natural disasters, flood damage may be covered by insurance companies in the same way as ordinary damage, in accordance with the principle of nation-wide solidarity.

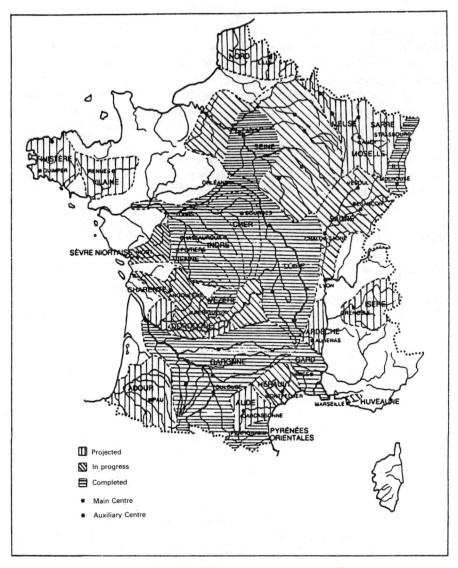

FIGURE 1 *Flood Warning Services networks*

Victims are eligible for compensation provided the following three conditions are satisfied:

– a state of natural disaster has been officially declared in an inter-ministerial decree;
– the damaged property is covered by a pre-existing insurance contract;
– the damage is declared to the insurance company within 10 days.

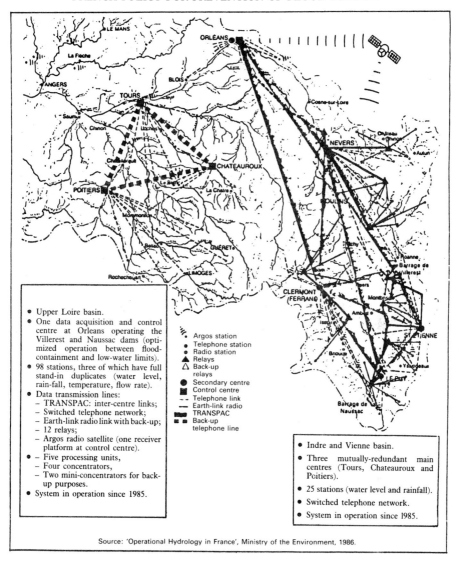

- Upper Loire basin.
- One data acquisition and control centre at Orleans operating the Villerest and Naussac dams (optimized operation between flood-containment and low-water limits).
- 98 stations, three of which have full stand-in duplicates (water level, rain-fall, temperature, flow rate).
- Data transmission lines:
  - TRANSPAC: inter-centre links;
  - Switched telephone network;
  - Earth-link radio link with back-up;
  - 12 relays;
  - Argos radio satellite (one receiver platform at control centre).
- – Five processing units,
  - Four concentrators,
  - Two mini-concentrators for back-up purposes.
- System in operation since 1985.

```
Argos station
Telephone station
Radio station
Relays
Back-up relays
Secondary centre
Control centre
Telephone link
Earth-link radio
TRANSPAC
Back-up telephone line
```

- Indre and Vienne basin.
- Three mutually-redundant main centres (Tours, Chateauroux and Poitiers).
- 25 stations (water level and rainfall).
- Switched telephone network.
- System in operation since 1985.

Source: 'Operational Hydrology in France', Ministry of the Environment, 1986.

FIGURE 2 *Loire basin: networks for transmitting hydrological data*

Further, certain types of crop damage are covered by the 'guarantee against agricultural disaster' system introduced in 1984.

Prevention: regulation of land use in flood-prone areas

The three-fold purpose of preventive measures is to assess the risk of flooding, to plan the use of land in flood-prone areas, and to educate the population affected as to the nature of the risk at hand. Such preventive measures mainly take the form of regulations issued by the State and local government bodies; continuing changes in these regulations reflect an increasing awareness of flood risks, especially as far as their social and economic impact is concerned.

The longest-standing State responsibility, and one that continues to grow in importance, is the exercise of a water planning policy. The main purpose of a water planning policy is to ensure the free circulation of water by controlling the works and constructions carried out in the beds of waterways, whether these lie within private or public property. State powers in this area reside with the *Préfets* of French *Départements* and are exercised by the Public Works Directorates, the Agriculture and Forestry Directorates and the Navigation Boards of the *Départements* concerned. A total staff of 620 is directly occupied with such tasks. The ministries concerned with water planning, in particular the Ministry of the Environment, have taken the necessary measures to clarify the complex legal basis for administrative action in this field. Further, the work of the bodies responsible for implementing the water planning policy has been revalued this year; additional resources have been allocated and clubs have been established to promote the exchange of ideas.

However, such flood-prevention measures will only be partially effective, as they are only applicable to the low-water channel. To extend the administration's powers to cover the entire flood plain, the decree-law of 30 October 1935 introduced plans and regulations applicable to flood-prone areas, restricting land use in order to ensure the free flow of water and conserve flood plains.

These plans and regulations establish public utility easements and are included as appendices to land use plans. Constructions, depots, plantations, enclosures and any other land use in areas classified as flood-prone must be declared to the administration, which has the right to impose special restrictions for the performance of works. At the present time, 3,300 km of valleys are covered by such plans.

The law of 13 July 1982 introduced Plans for the Reduction of Exposure to Foreseeable Natural Risks (PERs) (see Figure 3). The philosophy behind these plans is essentially to educate the population affected as to the nature of the risk at hand. This takes the form of individual preventive measures and restrictions to certain types of land use with a view to reducing the exposure to risk and consequently reducing the level of potential damage. Since 1984, 575 PERs covering flood risks have been ordered; documents have been drawn up for some 50 of these PERs, and the enquiry and approval stages have been completed for 12.

Protection Against Flood Waters

The protection of waterside land is theoretically incumbent on the owners of the land in question, whether private individuals or associations of land

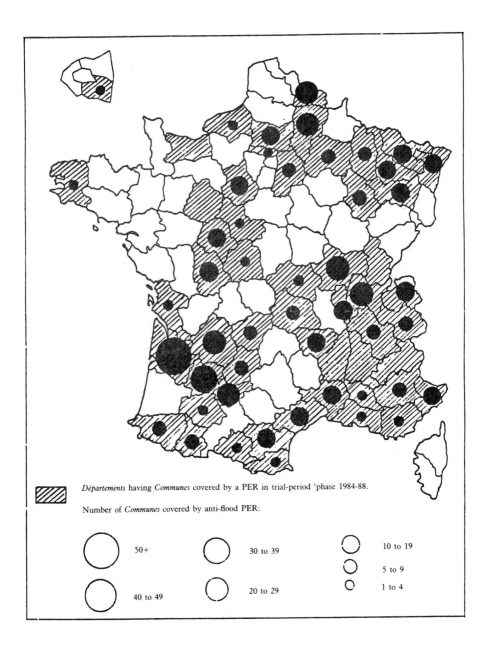

Départements having Communes covered by a PER in trial-period 'phase 1984-88.

Number of Communes covered by anti-flood PER:

50+

40 to 49

30 to 39

20 to 29

10 to 19

5 to 9

1 to 4

FIGURE 3 *Current coverage of Plans for the Reduction of Exposure to Risk (floods)*

owners. However, in view of the magnitude of the phenomena to be countered, the importance of the social and economic factors at stake and the high cost of flood protection works, this responsibility has been taken on by the State, followed by local government bodies.

Maintenance: the first priority

First on the list of measures to be promoted to reduce the risk of flooding is that of regular maintenance of rivers. On non-public watercourses, which account for some 250,000 km of rivers and streams; this task was traditionally performed by the riverside dwellers themselves. However with the decreasing use of these waters (watermills, commercial-scale fishing, use of dredged materials) waterway clearance was gradually falling into abandon.

For some 20 years now, local government bodies, mainly at *Commune* level, have taken on this responsibility by undertaking riverway restoration programs. There is an increasing awareness from these bodies of the need to limit the damaging consequences of flooding and to protect the local environment. From a technical point of view, active encouragement, notably in the form of a technical manual issued by the Directorate for Water and the Prevention of Pollution and Risks, is given to the use of 'natural' non-contaminating maintenance procedures.

Except for navigable canals and canalized rivers, regular maintenance of which is essential for the safety of navigation, the maintenance of public watercourses has also suffered deterioration owing to a decreasing State interest in waterways which no longer fulfill their original purpose. Despite the unfavourable economic context, the Ministry of the Environment, under whose responsibility these waterways fall, has this year been able to obtain new budgetary provisions which will enable recovery work to proceed on the major previously-navigable waterways. It should be possible to renew and extend these measures in the years to come, with particular attention to the development of water-borne tourism and water sports.

State intervention

The law of 7 March 1963 empowered local government bodies to undertake either individual or collective action to protect non-public riperian land wherever this was judged necessary for reasons of emergency or in the general public interest. The law of 10 July 1973 extended these powers to cover all works undertaken to protect against the encroachment of water. In passing, it should be mentioned that both of these laws make provision for charging part of the cost of works to the owners of the land in question wherever the land owner can be considered as gaining from the performance of such works. In practice, however, these provisions have rarely been applied.

An important State initiative aimed at attracting potential constructors takes the form of subsidies for works intended to protect agricultural land or inhabited areas. Such subsidies are awarded by the Ministry of Agriculture and the Ministry of the Environment respectively. Ministry of the Environment credits have seen a substantial increase in the last few years, and are expected to reach a total of some 160 million francs per year (overseas *Départements* and territories included). This corresponds to an overall investment rate of 500 million francs per year, which remains, however, well short of that needed to cover actual requirements, estimated at double this amount.

Flood protection methods

The flood protection techniques used are all well known: work on the bed of the watercourse to limit the risk of overflow (dyking, channelling, fixing banks and bed); construction of dams to contain flood waters. Decreasing construction costs coupled with the increasing reliability of design and construction techniques means that the construction of dams can be considered wherever the following criteria are satisfied:

- One or more potential sites must be available for storing the volume of water needed to ensure that the downstream water level remains below the required safety level;
- The construction cost, with due account taken of hydrological, geological and seismic factors, must remain compatible with the overall profitability of the operation;
- The project must be compatible with considerations of environmental conservation.

The Loire basin provides a good example of flood protection works: here, the Public Association for Planning of the Loire and its Tributaries, uniting all the local government bodies concerned, has undertaken a vast river training program which includes several large constructions.

Nowadays, the techniques for constructing dykes with shaped walls or with one impermeable wall plus drainage channel are also highly reliable; in those situations in which it is chosen not to adopt the solution of damming back waters during the flood period, the most frequently used technique is that of a combination of dyking and channelling. An example of this is the solution adopted for the Garonne at Agen and for the Haut-Rhône.

Special mention should be made of three factors which tend to increase the load to be taken by flood-water protection works. First, in rural areas, individual and collective land drainage operations and home modernization works decrease the surface water concentration time in the smaller catchment area tributaries, which in turn reduces the concentration time of peak flows during high-water period. Various studies are currently being undertaken with the purpose of quantifying these phenomena.

Secondly, in urban areas, the impermeabilization of soils has an even more marked effect. Striking examples of this are to be found in new towns under construction or expansion. Here, major construction programs are required to train rainwater flows within the settlement and to limit the amounts directed into rivers in order to ensure that flow capacities are not exceeded. Generally speaking, what is needed are rainwater catchment basins, dry wells, porous surfaces and other devices situated upstream or on the outskirts of settlements in order to delay flows and reduce surface water to a minimum. Difficulties concerning implantation into an already existing urban fabric can often hinder the implementation of essential installations of this kind.

The third factor is the existence, past or present, of extraction plants. By disturbing the bed load – an essential phenomenon for the morphological equilibrium of a watercourse – intensive extraction can de-stabilize the river, which will tend to shift by eroding its banks and bed until a new equilibrium is attained. For a number of years now, the Ministry of the Environment has adopted a firm policy limiting extraction from the low-water beds of watercourses under public ownership. These restrictions cover a total of 17,000 km of riverways and include the largest in the country. The first report on these measures was issued in 1985 and showed a very noticeable reduction in the volume of extractions, especially from the Loire, the Allier, the Dordogne, the Durance and the Garonne. Despite these measures, however, the consequences of past extraction operations will continue to be felt for a long time in the valleys concerned and will continue to hinder the implementation of efficient protection against flooding.

The effect on the environment: impact studies

In decisions to proceed with any kind of development work, a study of the effects on the environment should be taken as a determining factor. It is essential that environmental considerations be taken into account by the construction company right from the preliminary design stages, and that the same attention be paid to acquiring environmental data as to acquiring other physical data. Work such as re-channelling and dyking which is carried out along purely functional lines, i.e. on the basis of hydraulic criteria alone, will lend a wholly artificial aspect to the watercourse, to the detriment of the attractive natural surroundings of which it forms the centre. While the introduction of impact studies 10 years ago represented an important step forward as far as environmental issues go, it remains to be said that for such a study to carry its full weight, and for the environmental factor to become a true decision-influencing factor, analysis of the effects on the environment must be undertaken well in advance of the project itself.

This is particularly true with regard to flood-protection works, as can be clearly seen in the generally favourable reception given to those projects which have observed the above procedures, even where imperative issues regarding economic factors and the sensitivity of the riverian population have in fact played a determining role.

A New Risk-Prevention Policy

In July 1987, the French Parliament passed a law aimed at reorganizing its provisions for ensuring public safety, preventing forest fires and preventing natural disasters. This law, jointly put forward by the Minister for Home Affairs and the Minister for the Environment, marks an important step forward in the field of risk prevention, with particular emphasis on the prevention of flood risks. It seeks to promote a policy of education, information, prediction and improved land use.

Regarding public information for preventive purposes, citizens are recognized as having the right to be informed as to the nature of the natural and technological risks to which they might be subjected and as to the safety measures which have been or are to be taken to counter such risks. These safety measures, together with appropriate rescue procedures, are set out in conventional but restructured specialized emergency plans. Particular risks, or risks associated with particular constructions such a large dams, are dealt with in special action plans. These set out the measures to be taken in the vicinity of such constructions. After approval, these special action plans will replace the plans for alerting the public authorities which have applied up till now.

Regarding town planning, the risk of damage and the loss of life can be reduced by having all existing risks figure in the appropriate town planning documents. An example of this is the case of land use plans, in which mention must be made of foreseeable natural risks. Further, the plans for the reduction of exposure to risk are improved by their including the provisions figuring in the plans for flood-prone areas. Henceforth, plans for the reduction of exposure to risk will replace the plans for flood-prone areas as soon as they are released; all information regarding flood risk prevention and the measures to be taken in order not to exacerbate this risk will thus be available in a single document. The contents of the new plans for the reduction of exposure to risk will be specified in the near future in order to maintain the present release rate.

Finally, by modifying the provisions relating to the content of decisions affecting the authorization of constructions on watercourses, the 1987 law gives *Préfets* regulatory powers to prevent or cease any possible threats to public safety arising from the proximity of a dam. This law is intended to further an increased sense of responsibility from operating enterprises and administrations. To improve efficiency, steps will be taken to provide additional training and improve the resources available to the technical services responsible for enforcing the regulations concerning such constructions (see Appendix 3).

Actual application of the majority of the new provisions is still pending the preparation of texts which will determine their modes of enforcement. However, the authorities responsible for enforcement are already actively involved in communicating the nature of the objectives sought to those concerned (local government bodies, construction companies, administrations, users). Thus, the public bodies in France are pursuing an active policy aimed at reducing flood risks. Design offices, laboratories and companies in France act as efficient instruments in the application of this policy. These enterprises are in a position to make their know-how and capabilities available to third-world countries, all too often subject to problems in this area which are much graver than those suffered by ourselves and which must imperatively be resolved if development is to be allowed to proceed and if national survival is to be guaranteed.

BIBLIOGRAPHY

JACQ, A. and ROCHE, P.A. (1986) Operational hydrology in France. Ministry of the Environment, Paris

ROCHE, P.A., et al. (1987) Guide to flood forecasting. Société Hydrotechnique de France, Paris

MINISTÈRE DE L'ENVIRONNEMENT – AGENCE DE BASSIN (1985) Watercourse Maintenance. Technical manual from the Directorate for Pollution Prevention, no. 14

MINISTÈRE DE L'ENVIRONNEMENT (1988) Floods – practical guide, Paris. (1988) Catalogue of flood prevention measures (provisional edition), DRM. (in press) Cartography of plans for the reduction of exposure to flood risk, DRM

GARRY, G. and LE MOIGNE, M. (1985) Photo-interpretation and cartography of flood prone areas. DDP-DRM-STU, Paris

APPENDIX 1

The Grand-Bornand Disaster

On 14 July 1987, as the result of an exceptionally violent and long-lasting storm falling on already waterlogged ground, the river Borne overflowed causing the death of 23 people on a riverside camping site. Apart from its tragic aspect, this disaster also gave rise to controversy regarding the concept of flood forecasting, since preliminary studies for a plan for the reduction of exposure to risks (PER) had already been carried out in the Commune in question. A commission of three inspector-generals was instructed to hold an administrative enquiry into the circumstances and causes of the disaster and come up with general recommendations for ensuring the safety of campers on sites in mountain areas. The commission first determined that there was no question of blame or negligence and that the disaster was essentially the result of an exceptional natural phenomenon, though the location of a camping site in a flood-prone area could be considered unwise. It then went on to make the following five proposals:

(1) *A list should be drawn up of all camping/caravan sites exposed to natural risks.* This list, currently under preparation, shows more than 3,000 sites to be affected.

(2) *The preparation of PERs should be accelerated and the technical studies involved should be simplified.* Since the summer of 1987, this recommendation is being followed. PERs are to cover a total of 660 *Communes*; as yet, 45 (only) of these PERs have been issued and 15 have been approved. The objective is to complete half the total number by the second quarter of 1988. Methodologies are being simplified, but efforts in this direction are limited by the need to reach very precise decisions.

(3) *Legal status should be afforded to the hazards map, and R111-3 risk perimeters should be established in those locations where no PERs have been ordered.* The hazards map must be communicated for study by the elected officials. However, for this map to have legal status, it must necessarily be accompanied by appropriate prescriptive orders, which means that there is no advantage to be obtained with respect to a PIG or R111-3.

(4) *The Préfet should be able to refer a mayor's decision to the administrative court, suspending the execution of such for reasons of public safety.* This measure is under study.

(5) *The procedures for authorizing camping and caravan sites should take fuller account of natural risks.* It will be possible to implement this recommendation when considering the authorization of future sites. As regards currently existing sites, economic and tourist-trade requirements would appear to conflict with public safety obligations. In this matter, the Ministry of the Environment considers that, in the most important cases, the risk of legal dispute is preferable to that of a further disaster.

Information for use by *Préfets* in this area is currently under preparation.

APPENDIX 2

Example of a flood warning network: the Loire basin

The Loire basin accounts for a total surface area of some 115,000 km^2. An analysis of the task at hand in the Loire basin brings to light the existence of two systems calling for two very different data acquisition strategies. The damage liable to be caused by flooding of certain tributaries of the middle stretches of the Loire (the Indre and the Vienne) is significant but limited in magnitude. The flood warning system used here is simple in design and inexpensive to run. In contrast to this, the upper reaches are prone to heavy and rapid flooding; here, multiple constructions are required along the watercourse. The basin of the upper Loire, then, has been equipped with one of the most efficient flood warning networks in France. Five secondary sub-stations are needed to cover the expanse of the basin of the upper Loire. The flood warning network must fulfill three basic purposes:

– Control the Villerest and Naussac dams, particularly at flood-containment and low-water levels;
– Monitor low-water levels;
– Handle flood forecast and flood warning operations (performed in secondary stations)

To satisfy these objectives, high-reliability links with monitoring stations are essential. For this reason, the solution adopted for data transfer combines a backed-up radio-electric network, ARGOS satellite facilities and the switched telephone network. The system as a whole is coordinated by the data acquisition and control centre, which operates the dam control centre. Stations are equipped with earth-link radio transmitters and are polled from four automatic data concentrators located upstream of the basin near the secondary centres. Each is allocated a three-minute timespan and polling may be performed at four-hourly, two-hourly, one-hourly or half-hourly intervals.

Data are transmitted via the TRANSPAC network to the processing units at the different centres. Each centre filters the relevant data and implements forecast models. The control centre at Orleans queries all telephone stations, and retrieves data from all ARGOS stations using its direct reception station. Each secondary centre calls those telephone stations in its area. All data exchange between centres is via the TRANSPAC network.

Provision has been made for a number of back-up procedures:

– Each station may be reached by two relays;
– Each concentrator may be replaced by another;
– A series of intermediary radio relays will take over data transfer in the event of a failure on TRANSPAC;
– Full stand-in duplicates exist for three key stations.

With this high degree of redundancy, the control centre is able to permanently monitor the flow of the Loire and optimize dam operation.

A complex network of this type is justified when substantial economic factors are at stake (see Figure 2).

APPENDIX 3

French regulations concerning dams

State regulations concerning water-retaining structures were revised by the relevant authorities following the Malpasset dam disaster of 1959. The following three legal provisions ensued:

(1) For dams more than 20 m high, the Permanent Technical Committee for Dams must be consulted at the preliminary project stage preceding

construction work and before all reinforcement work is undertaken. This Permanent Technical Committee was set up in 1966 and is composed of qualified members of the administration and representatives from private industry.

(2) All dams liable to represent a risk to public safety are subject to inspection on a regular basis. The inter-ministerial order of 14 August 1970 set out the respective duties of the dam operator and the State regarding the regular inspection of constructions, starting from initial dam-filling and covering any special revisions which might be deemed necessary, especially in the case of older dams. Primary responsibility for the upkeep and maintenance of the constructin lies with the owner or his concession-holder.

(3) Emergency plans are required for major dams. The decree of 16 May 1988 requires that hydraulic constructions with dams of more than 20 m high and retention capacities of greater than 15 million m^3 be covered by emergency plans specifying the measures to be taken for alerting the authorities. Some 88 operational dams are covered by this provision; 66 of these are hydroelectric dams run by the French Electricity Board (EDF).

Regulatory measures, public safety measures, and the role of the Permanent Technical Committee for Dams are to be reinforced by the texts enforcing law no. 87–565 of 22 July 1987, concerning the organization of public safety, prevention of forest fires and prevention of major risks.

Some Problems in Flood Disaster Prevention in Developing Countries

—

MASAYUKI WATANABE

United Nations Office of the Disaster Relief Co-ordinator, UNDRO, Palais des Nations, Geneva, Switzerland

Introduction

1987 will be remembered as a year of problems due to natural disasters which occurred in many parts of the world. It included, in addition to the lingering drought in the African continent, a tidal wave which submerged the Maldives in April. An horrific fire broke out in northern China; yet another disaster was the serious shortage of monsoon rain resulting in drought in India. Meanwhile, neighbouring Bangladesh was fighting an endless war against excess water supply from the regions beyond the border and within her territory. Twenty-one million people were affected and more than 50% of the land was submerged there.

Just after the heat wave which hit Greece seriously, neighbouring European countries were struggling against floods and landslides due to unusual heavy and sustained rainfall and wind-storms. The Philippines and Vietnam, and island countries in the South Pacific and South-West Indian Ocean were seriously affected by typhoons and cyclones.

Every country suffering from the disasters described above revealed more or less vulnerable points in terms of disaster management.

UNDRO's Response to the Disaster Situation

Bangladesh flood

No sooner had UNDRO received the official request from the Government than a relief co-ordination officer left for the disaster stricken country to assess the damage and estimate the relief requirements of the victims. His reports from the field formed the basis both of the United Nations Disaster Relief Co-ordinator's appeal and of the subsequent situation reports which were sent out to all major relief donors of the world.

These reports provide a graphic picture of the destruction in the regions and the latest relief requirements of the victims in accordance with the latest development of the rescue and relief works in the stricken areas.

The UNDRO officer, in co-operation with the Resident Representative of the United Nations Development Programme, also co-ordinates the activities of relief groups which include representatives of potential donor governments, inter-governmental organizations, non-governmental organizations, the United Nations system and other organizations involved in humanitarian relief in the field so as to ensure timely and efficient delivery of relief goods and materials and humanitarian services, and to avoid duplication and loss.

The Disaster Relief Co-ordinator, Mr Essaafi, visited the flood-stricken regions in response to the invitation of the President and the request from the Secretary General of the United Nations. His mission was followed by the damage assessment mission organized jointly by UNDRO and UNDP.

Emergency contributions received in response to UNDRO's appeal amounted to over US$72 million, and covered all relief needs.

Other disasters in which UNDRO was involved

Similar operations had been undertaken for the cases which took place one after another due to cyclones, volcanic activity, earthquakes, wild fires, locust infestations, landslides, tsunamis, etc. The cyclone disasters in Bangladesh, the Philippines, the volcanic mudflows in Colombia in 1985, the earthquake in Mexico and El Salvador in 1987 are typical examples.

Special efforts have also been made to cope with creeping disasters such as food shortage in Africa and the man-induced disasters in Sudan, Mozambique, Lebanon and South Yemen.

Disaster Prevention and Preparedness

In addition to co-ordinating relief efforts, UNDRO assists in raising the standard of pre-disaster planning and preparedness including disaster assessment and relief management capability in disaster-prone countries, as well as promoting the study, prevention, control and prediction of natural disasters.

Disaster prevention and preparedness consists of a wide range of measures, both long and short term, designed to save lives and limit the amount of damage that might otherwise be caused by the disaster. Prevention is concerned with long-term policies and programs to prevent or eliminate the occurrence of disasters. The corresponding measures are taken in such fields as legislation, physical and urban planning, public works and building.

Preparedness is not limited only to the short-term measures which are taken during a warning period before the impact of a disaster event: it must be supported by legislation, be concerned with operational planning, education and training of the population at large and the technical training of those who will be required to help in a relief operation: stockpiling of supplies and emergency funding arrangements.

In this context, UNDRO has recently been implementing disaster mitigation projects in the Southeast Asian, Balkan, Caribbean, Latin American and Mediterranean regions, providing workshops and seminars, technical assistance in dealing with earthquake, volcanic and tsunami risks, and preparing technical manuals and monographs on mudflows and public education.

Training is another important factor for sensitizing officials in disaster-prone developing countries to the need for prevention and preparedness. The Asian Disaster Preparedness Centre in Bangkok, established as a result of an UNDRO initiative, offers courses in disaster prevention and preparedness which attract officials from all over the region which constitutes the South-East and West Asian and the Pacific regions.

Problems Associated with Flood Disasters

Human failure

One of UNDRO's mission reports on flood disaster due to tropical cyclone revealed that any investment in reconstruction should be based on a solid understanding of the problems involved. With regard to the reconstruction for the dyke of the river A in country B, for example, to rebuild the same type of dyke in the same place, or at least under similar conditions, would probably be a waste of effort and would not prevent further disasters.

Between 1928, when the original dyke was built, and the present time the entire river system has changed: the course of the river channel has altered, runoff problems and silting have increased due to unsatisfactory land-use practices and controls, etc. To arrive at effective flood prevention the river, the watershed areas around it and upstream development in the river basin area have to be taken into account, as would future planned development in the river basin area. So-called emergency aid to reconstruction in the form requested by the government could possibly be counter-productive.

Another report pointed out that the majority of the damage results from human failure rather than from the effects of nature; buildings badly located, dykes and bridges poorly built, etc. These human failures are frequently due to lack of financial means but almost as often to lack of knowledge.

Causes of flood disasters

A report on a flood disaster illustrates the problems as follows:

'The population is growing and people are increasingly being forced to build dwellings and food stores in highly vulnerable situations. The flood waters to a great extent rise from beyond the national boundaries. They are increasing in volume. Positive environmental management is still in a very early stage of development over the borders and within the country.

Flood protection is normally provided by building earth embankments, or by simply-constructed walls along a river channel. Road and railway embankments may act as unintended flood retention embankments or drainage hindrances depending on their locations and the sluices and culverts constructed along them.

Embankment construction and maintenance is complicated by the unstable nature of the rivers and their high waterlevel. Embankment are frequently eroded when rivers move course or when they are close to the main channel. Polders can only be constructed if they incorporate an adequate drainage system, even then heavy rainfall may cause flooding when outside river levels are high' (UNDRO, 1987).

Major causes of the disaster identified in the above report are as follows:

(a) *Overtopping of embankment*
Overtopping of embankment may be acceptable if the frequency is not too high and the embankment is not swept away in the flooding process;

(b) *Public cuts*
Public cuts may indicate a need for drainage system improvement, or education and co-ordination, to prevent conflicts of interest and uniformed actions causing greater losses;

(c) *Breaches*
Breaches are likely to indicate poor construction or design and poor management;

(d) *Erosion*
Erosion may be inevitable along major rivers, but afforestation in front of the embankment and brick revetments held in by wire appeared quite effective in protecting limited sections of embankment.

(e) *Partial completion*
Partial completion between two construction seasons exposes an earthwork structure to damage at the in-between flood season.

(f) *Maintenance failure*
Engineering structures that failed had already damaged, cracked and pot-holed surfaces and the new flood accelerated the maintenance failure.

Forecasting and Warning

Reliable forecasting and easily understandable warning information with sufficient lead time are of vital importance for evacuation. These are the final measures which can assist in survival.

This issue constitutes three aspects: data acquisition and processing; warning dissemination; and public response to the warning provided.

The following description which appeared in an official report provides vivid accounts of the existing situation and problems over the systems for data acquisition:

'Although more numerous than existing river gauge stations, these new rain gauge points are also spaced far apart with each observation point generally covering over 260 km^2 in both the upper and lower catchments. Very few stations are equipped with self-recording devices and facilities for communicating through carrier telephones. Local observers come to the stations to record rainfall data, then go to the nearest post office to book a telephone call.

Most of the messages are not received, because there are invariable failures in the land-line telegram channels during periods of heavy rainfall, thunderstorms, and power cutoffs' (Government of West Bengal, 1981).

With regard to data processing, and putting technological problems aside, a sense of responsibility is the basis of all disaster prevention and relief activities. The following articles reveal what was behind the tragedy due to a tropical cyclone:

'One particular satellite picture was printed with the date and time on it. A careful observation reveals that the date was put on after erasing some other date and looked different from others and words corresponding to typical computer printing. Was the picture received at some other day before? It was reported in a section of the press, after the cyclone, that the picture was published in a manner as if it was received one day after.

Certain quarters also view that the radio forecast system failed to work properly. It has been reported afterwards that the meteorological authority of the neighbouring country conveyed to its counterpart news about the cyclone two days before its occurrence. It may be mentioned that both the agencies normally exchange relevant information under a bilateral agreement. The information on the cyclone which was proceeding at a speed of 100–120 km/hour was provided beforehand. It means that the authority had prior information of the imminent danger to an extent of at least 48 hours. The satellite picture further confirms it.

The Central Radio went off after closing its normal programmes at the scheduled time on the day of the tragedy. It was expected that repeated radio forecasts with cautionary warnings would be made until the danger was over. In such an emergency situation, the behaviour of the radio station was rather surprising. The zenith of punctuality indeed!' (Ahmed, 1985).

'A lack of knowledge of upstream flood mitigation schemes, e.g. channel enlargement or construction of entirely new channels or embankments, limited the villagers' appreciation of the flood possibilities that may occur. In other words, knowledge and experience of previous floods ('folk memory') could not provide a reliable basis for response to a flood warning unless adequate information about flood inverventions undertaken upstream were given to the occupants of downstream areas' (Government of West Bengal, 1981).

With regard to warning dissemination the following record of interviews and surveys conducted in West Bengal show where the problems lie:

'Most of the families who are victims of floods have complained that there is no warning system. They only know about the flood when they are surrounded by flood waters. Only 10%–15% of the families are informed in time' (Sikander, 1983).

'Officially the police stations, since they are equipped with radio transmitters, represent the level of the civic hierarchy responsible for issuing flood warning messages. These are usually disseminated by police with loudspeakers touring the area within their jurisdiction in jeeps. The usefulness of police in disseminating warnings depends largely on the distance they must travel from their stations to remote villages, prevailing weather conditions, and the concern of individual police officers in matters of floods. Nearly 80% of the villages are 5 km from the nearest police station and more than 50% are over 5 km from an all-weather road.

During the early morning hours of 4 October the Teesta Bridge gauge passed the extreme warning stage level. Although flood warning rules specified that messages were then to be conveyed as 'Danger level' and 'Extreme danger level', the message used weaker terms 'Warning level' and 'Extreme warning level'. By 8.30 am the message had been received by the radio station but was not communicated to the public. Apparently, the advisory official was reluctant to issue the flood alert because there was no imminent 'danger' ' (Government of West Bengal, 1981).

'According to the Central Radio officials, the technical words used in bulletins are usually not familiar to the program executives, handicapping their efforts to explain to listeners. Use of technical words should, therefore, be avoided' (Sinha and Avrani, 1984).

'More emphasis needs to be given to the dissemination and interpretation of official warning messages in the local language; and similarly to public education measures designed to improve community awareness of the risks and preparedness measures the authorities in Dhaka issued repeated warnings of the impending disaster, and accurately forecast the path of the cyclone, including the final alert No. 9, which meant immediate evacuation from the danger zone. But many inhabitants on the islands straddling the path of the cyclone missed the alert for the simple reason that they did not own or have access to radios' (Ranganathan, 1985).

Concerning public response to warnings, the problems to be tackled are not so simple. Social, economic and religious problems lie behind disasters:

'Villagers tend to leave their homes after flood warnings only as a last resort because they believe that thieves in boats will steal their belongings. They cannot afford to leave behind their few household possessions and tools' (Mathur, 1981).

'The mere conveyance of warning messages is not enough to protect lives and property. Administrative officials, as well as flood-affected residents, need to have information on what they should do after they receive a flood warning' (Government of West Bengal, 1981).

Damage and Loss Assessment

Although the magnitude and nature of floods varies from place to place and in accordance with the changes and degradation taking place in catchment basins, most of the flood disasters are in fact a recurrence of similar events in the past.

The key to success in disaster mitigation, in most cases, is to learn lessons from past events. The data and information on the real causes and dynamic processes of partial and total damage and destruction offer the key to proper measures for disaster mitigation and damage reduction.

According to the report on the flood disaster in Bangladesh in 1987, damage is classified as 'full', defined as a 'public cut' breach, or washing away of embankment or structure or 'partial': defined as overtopping of embankments, some erosion of embankment sides, or damage to a structure which can be repaired. Damage is costed by responsible engineers based on annually revised scales of contractors' charges. The basis of estimating is inevitably rough and probably uneven among the officers in charge (UNDRO, 1987).

The data acquired based on unclear and rough terminology, however, are not the kind of figures on which rehabilitation or reconstruction programming or rural and resources development programmes can be based.

A similar statement can be made regarding failure analysis. Good design and proper construction and maintenance practice become better through the understanding of the patterns and processes of structural failure and environmental changes which take place in adjacent reaches of river channels.

This concept was presented by UNDRO to the Disaster Prevention and Preparedness Component of the Tropical Cyclone Committee for the Typhoon region and the Panel on Tropical Cyclones region in 1986 (ESCAP/WMO, 1986; WMO/ESCAP, 1986).

Some Hazard Events which give Serious Impacts to River Systems

Vulnerability analysis in riparian regions

The natural disasters of March 1987 had a grave impact on social and economic development in Ecuador.

The initial cause of the disaster were two earthquakes with magnitudes of 6.0 and 6.8 on the Richter scale. The earthquakes triggered hyge landslides, yielding a vast amount of debris material which formed ephemeral landslide dams with a large volume of water retained behind them. Landslide dams breached abruptly creating hydraulic bores; these in turn contributed to major damage both to the Trans-Ecuadorian oil pipeline, used to transport oil from and Amazon region to the refineries and export centres on the Pacific coast, and to the highway connecting the eastern provinces with the rest of the country.

Floods carried away thousands of head of livestock and covered grazing land with sediment.

This disaster raises an essential point in vulnerability analysis in riparian regions and suggests that closer interdisciplinary co-operation is required for water-related multiple hazard prevention.

Flash flood

The term 'flash flood' has come to be used frequently in news reports of consequences of tropical cyclones and torrential and sustained rainfall all over the world.

Judging from the mode and the nature of discharge, most of the so-called 'flash floods' are hyperconcentrated flows which result in irreversible topographic changes and complete destruction of alluvial cones and fans.

Conventional measures for river training are of no use in coping with disasters due to this sort of flow, yet limited knowkedge and technology have been developed and applied in disaster prevention, watershed management and water resources development in developing countries.

History shows that any kind of development activity undertaken to support growing population in a catchment basin inevitably results in a remarkable increase in sediment yield and transportation unless well planned watershed management programmes are carried out.

Much more emphasis, therefore, should be put on the problems of excess yield and transportation of sediment, on the social background, and on measures for coping with the problems associated with sediment.

Hydraulic Monitoring

Networks and training for hydrologic monitoring have been established and organized fairly successfully as a result of international, regional and bilateral co-operation programmes promoted through IHP, TCP, etc. Hydraulic monitoring practice and education and training, however, are very poor.

Numerous water resources development programmes and disaster prevention programmes have failed or are on the brink of collapse. Old villages and towns are on the verge of destruction because of the shifts of channel course or riverbed changes.

There is an urgent need to stimulate hydraulic monitoring practice, education, training and public information programmes in developing countries.

More and Efficient Allocation of Resources for Flood Disaster Reduction and River Basin Management

Flood can be identified as the principal disaster agent in a number of events amongst drought, flood, civil strife, tropical cyclone and earthquake.

Althouth it is the agent in fifth place in terms of the average death toll, the fastest increase appears to be in floods: both in terms of frequency and the number of people affected during the period from 1960 to 1980. This means that efforts for flood disaster prevention made so far do not meet, and are far behind, the real needs in our world.

UNDRO's report on the Bangladesh flood revealed dramatically that a major problem is the large 'refugee' population (10,000 households in this district) who made cuts into the embankment, where they live, to build themselves shelters.

Measures Against Hyperconcentrated Flows

Characteristics of hyperconcentrated flows

A flow which is heavily laden with sediments is referred to as a hyperconcentrated flow. Hyperconcentrated flows are also referred to as mudflows or debris flow according to their particle size distribution. Hyperconcentrated flows are often referred to as volcanic mudflow or lahar in the case of the discharge in active volcanic regions. Particle size distribution and material source are, however, not necessarily decisive factors in defining the mode of a flow.

The mode of flow is classified as shown in Figure 1 but, at the same time, can be expressed by Equation (6) according to the distribution of share stress in flow.

$$\frac{\varrho u_{\star}c^2}{(\sigma - \varrho)gd} = 0.034 \cos \theta \left[\tan \phi - \frac{\sigma}{(\sigma - \varrho)} \tan \theta \right] \times 10^{0.32(d/h)} \qquad (1)$$

$$\tan \theta = \frac{C_{\star} (\sigma - \varrho)}{C_{\star}(\sigma - \varrho) + \varrho(1 + h/d)} \tan \phi \qquad (2)$$

$$\tan \theta = \frac{C_{\star}(\sigma - \varrho)}{C_{\star}(\sigma - \varrho) + \varrho(1 + k^{-1})} \tan \phi \qquad (3)$$

$$\tan \theta = \frac{C_{\star}(\sigma - \varrho)}{C_{\star}(\sigma - \varrho) + \varrho} \tan \phi \qquad (4)$$

$$\theta = \phi \qquad (5)$$

where $u_{\star}c$ is the critical shear velocity, d is the diameter of grain, g is the gravitational acceleration, ϕ is the internal friction angle of grain, θ is the channel slope, h is the flow depth, $C_{\star}$ is the grain concentration in volume in the static channel bed, k is the experimental constant, σ and ϱ are the density of grains and fluid respectively.

$$\tau = C_0 + \tau_y + \mu \left(\frac{du}{dz} \right) + A(1 - e^2) \sigma \frac{1}{b} \left| \frac{du}{dz} \right| \left(\frac{du}{dz} \right) + \varrho \overline{u'u'} \qquad (6)$$

where τ is the share stress, C_0 is the viscosity of fluid, τ_y is the stress due to friction among particles, μ is the dynamic viscosity, z is the flow depth, u and u' are the velocity and mean velocity, respectively.

The differences between volcanic mudflow and mudflows with other origins

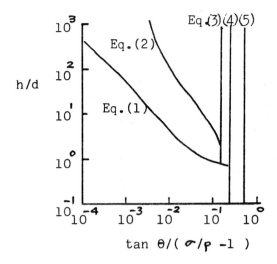

FIGURE 1 *Mode of sediment transportation (Ashida et al., 1981)*

stems from the magnitude of flow in terms of sediment discharge.

The sediment discharge in the case of the mudflow disasters due to volcanic eruption and glacier-melt is in many cases far greater than that of the mudflows due to torrential rainfall as shown in Table 1.

Simulation of volcanic mudflow on an alluvial cone and an alluvial plain

(1) *Simulation model for a mudflow in mountain reaches* (Mizuyama et al., 1988)

The dynamic equation of a volcanic mudflow in the reaches of mountain torrent may be expressed as

$$\frac{\partial q}{\partial t} + \beta \frac{\partial(qu)}{\partial x} = -gh \frac{\partial H}{\partial x} - \frac{F}{\gamma} \tag{7}$$

where q is the discharge per unit width, u is the mean velocity, H is the flow height (flow depth + bed elevation), r is the unit weight of water, F is the frictional resistance and is expressed by

$$F = gn^2 \frac{u|u|}{h^{\frac{1}{3}}} \tag{8}$$

where g is the gravitational acceleration of gravity, n is the Manning's roughness coefficient.

The continuity equation of water is

$$\frac{\partial h}{\partial t} + \frac{\partial q}{\partial x} = 0 \tag{9}$$

TABLE 1 *Volcanic Mudflows and their Magnitude*

Case	Total Discharge	Peak Discharge
Mt. St. Helens (eruption, 1980)	76,000 × 1000 m3	3,300 m 3/s
Nevado sel Ruiz (glacier melt, 1985)	43,300	29,000
Mt. Tokachi (eruption, 1925)	13,300	1,300
Sakurajima (rain, 1981)	100 – 300	100 – 500
Semeru (rain, 1981)	1,000	800
Usu (rain, 1981)	400	100

while that of sediment is

$$C_* \frac{\partial z}{\partial t} + \frac{\partial qB}{\partial x} = 0 \tag{10}$$

where h is the flow depth, z is the sedimentation depth, C_* is the volumetric concentration of the materials of the channel bed, qB is the sediment discharge per unit width of the flow.

qB can be calculated by the Mayer-Peter-Muller formula.

(2) *Simulation model for a mudflow on a plain*
In the case of a mudflow which runs on a plain, the dynamic equation is expressed as

$$\frac{\partial M}{\partial t} + \beta \frac{\partial(Mu)}{\partial x} + \beta \frac{\partial(Mv)}{\partial y} = -gh \frac{\partial H}{\partial x} - \frac{Fx}{\gamma} \tag{11}$$

and

$$\frac{\partial N}{\partial t} + \beta \frac{\partial(Nu)}{\partial x} + \beta \frac{\partial(Nv)}{\partial y} = -gh \frac{\partial H}{\partial y} - \frac{Fy}{\gamma} \tag{12}$$

where M is the discharge per unit width in the x direction, N is the discharge per unit width in the y direction, u is the mean velocity in the x direction, v is the mean velocity in the y direction, H is the flow height (flow depth + bed elevation), r is the density of the mudflow, Fx and Fy are the frictional resistance in the x and y direction respectively.

Fx and Fy are expressed as

$$Fx = gn^2 \frac{u\sqrt{u^2 + v^2}}{h^{\frac{1}{3}}}, \quad Fy = gn^2 \frac{v\sqrt{u^2 + v^2}}{h^{\frac{1}{3}}} \qquad (13)$$

The continuity equation of water is expressed as

$$\frac{\partial h}{\partial t} + \frac{\partial M}{\partial x} + \frac{\partial N}{\partial y} = 0 \qquad (14)$$

and that of sediment is expressed as

$$C_* \frac{\partial z}{\partial t} + \frac{\partial}{\partial x}(C_M M) + \frac{\partial}{\partial y}(C_N N) = 0 \qquad (15)$$

where C_M is the sediment concentration of the flow in the x direction, C_N is the sediment concentration of the flow in the y direction.

The sediment concentration can be shown to be either Figure 2 or equation (16)

$$C = \frac{12n\sqrt{g}}{sh^{\frac{1}{6}}} i \qquad (16)$$

where $s = (\sigma/\varrho - 1)$.

Figures 3(a) and (b) show the results of the simulation by hydraulic model test and the numeric model respectively on the case of the Hachiemon alluvial cone in Japan. It is observed that each of the results represents the real events fairly well (Yazawa et al., 1986).

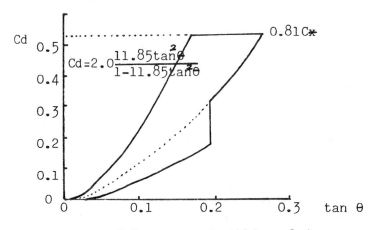

FIGURE 2 *Sediment concentration (debris-type flow)*

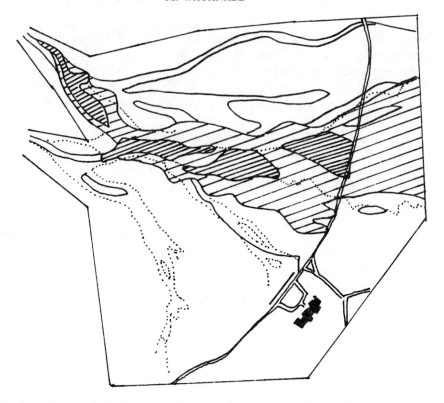

FIGURE 3(a) *Debris Flow Stricken Area . . . comparison between the actual situation and hydraulic model test.* ▨ *, sedimentation more than 1 m;* ▱ *, sedimentation less than 1 m;* ▭ *, surface flow;* ⌐ ⌐ *, actual hazard area*

Figures 4(a) and (b) show the affected area in the case of Mt Tokachi mudflow disaster in 1925 and in time sequence computed by the numeric model respectively. It apparently represents the real event in 1925 (PWRI). Simulation by either hydraulic model test or computer-aided numeric model can thus be powerful tools for hazard area delineation, planning of structural measures, identification of evacuation spots and early warning.

Structural measures against hyperconcentrated flows

(1) Dehydration of the flow

Past disasters clearly show that structural measures which are commonly used for river improvement work on an alluvial plain cannot be employed as

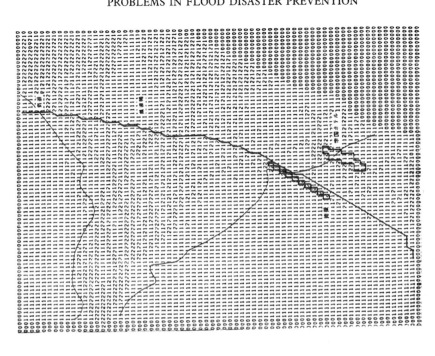

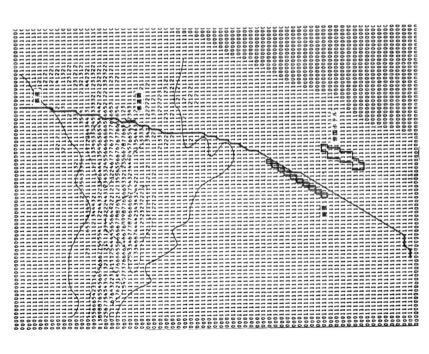

FIGURE 3(b) (i) Sedimentation area obtained by computer simulation; (ii) flooded area obtained by computer simulation

FIGURE 4(a) *Mudflow stricken area, 1925, Furano river, Japan*

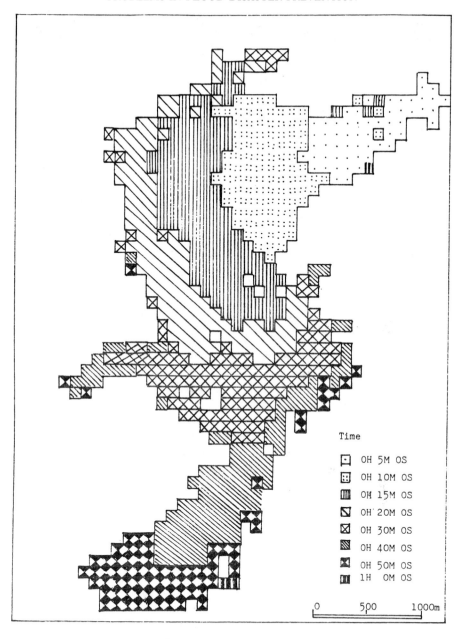

FIGURE 4(b) *Mudflow stricken area obtained by computer simulation*

FIGURE 5 *Screen-type debris flow breaker designed by the author*

measures against hyperconcentrated flows. Remarkable changes in channel course, longitudinal profile and sediment concentration, together with the destructive impact force of a flow, endanger almost all the engineering structures on alluvial cones and fans.

Disasters due to hyperconcentrated flow are characterized first by complete destruction due to impact force; secondly, by the magnitude of topographic changes and damage due to sedimentation; thirdly, by the extreme difficulty of evacuation upon the occurrence of the flow. It is almost impossible, and not always advantageous, to prevent hyperconcentrated flows from initiating.

From the engineering point of view, it is therefore feasible to take the following two measures:

– to remove large-sized material as much as possible;
– to make the sediment concentration as low as possible.

The lower the sediment concentration is, the easier and more feasible it is to control the flow by means of ordinary river control works.

Kinetic energy carried by large-sized particles can be decreased by the use of sabo dams which create some reaches with gentle bed slope in their reservoirs. Kinetic energy can also be decreased by dehydration: a horizontal screen set up on the channel bed has been proved to be effective for dehydration. The design elements of the screen-type debris breaker are given based on the hydraulic tests and field observation.

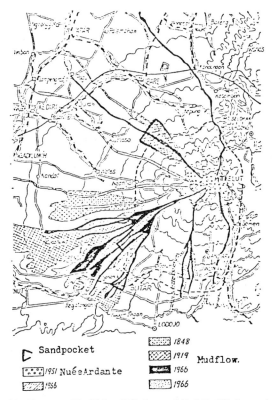

Sandpocket

[:::::] 1848
[////] 1919 Mudflow.
[===] 1966
[:::::] 1966

/951 NuéeArdante
/756

FIGURE 6 *Mudflows at Mt Kelut (Ministry of Public Works and Power, 1969)*

(2) Fan segment shift

Disasters due to lahar are characterized by the wide spread of the materials with wide variety of particle size.

The damaged area may increase depending on the location of the intersection point that shifts to and fro indefinitely, in accordance with the runoff and its sediment concentration. The key to success is to make the sediment concentration lower by making the most effective use of sand pockets. The size of the sand pockets may depend on the mode of the sediment movement, discharge and the location of the pocket. The sand pocket as shown in Figure 6 is ideal if enough area is available.

From the geomorphological point of view damaged areas are regarded as those where the fan segment is being formed, while disaster-prone area are those which will be affected once the formation elsewhere is completed.

The location and size of the sand pocket depend on the mode of the sediment movement, magnitude and frequency of sediment discharge and the condition of human use of the fan.

The land that compares with the damaged area or with the area extended to

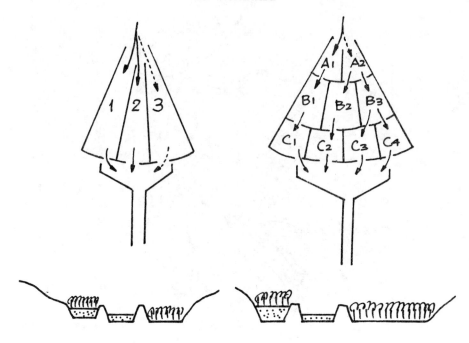

FIGURE 7 *Concept of the fan-segment shift as a measure for fan formation control*

the extent of five- to 10-fold at the minimum is desirable as a fan segment.

The sand pocket is enclosed in an embankment made of the materials available on the site, so that the height of the embankment is limited, rather low, and the life of the sand pocket not long.

New land should be secured for the next fan, and the flow will be guided to the new fan segment as soon as the one currently used is filled, as shown in Figure 7 (Watanabe, 1987). This operation would be continued as long as sediment transportation continues.

The inner part of the sand pocket is too disadvantageous and dangerous to be used for cultivation and housing but the one already filled-in can be used for cultivation. In this way, every corner of the alluvial fan currently under the threat of debris flows and lahars could be used efficiently without any fear of disasters if the corner is used as a sand pocket and the sand pocket is shifted in a certain systematic order as shown in Figure 8 (Watanabe, 1987).

The fan-segment shift method is an ideal measure for artificial fan formation process control, and is feasible as a preventive measure against hyperconcentrated flows.

(3) *Some technical points on the fan-segment shift method*
Ordinary river training works such as embankment, groin, ground sill and

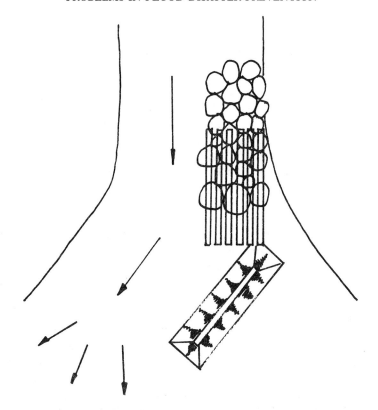

FIGURE 8 *Diversion of debris flow taking advantage of the dehydrate effect of a screen debris flow breaker set at the fan head*

gate can be used for diversion at the inlet of the sand pocket if the flow contains less and finer sediment.

But, in order to divert debris flows and lahars, ordinary river training works are of no use, while solid structures such as dams and guide walls are not really feasible, considering the cost for initial construction and maintenance.

The measures for diversion should be tough enough to resist the huge impact of debris flows and stable enough to cope with the large fluctuations of the river bed downstream. The best way to avoid the collapse of the facility for diversion is to make the best use of the mechanism of debris-lobe formation, and dehydration method by means of screen-type debris breaker can be employed as the artificial debris-lobe formation measure.

A screen-type debris breaker set up upstream, closer to the inlet, creates artificial debris lobes which can guide the direction of following flows, as shown in Figure 8. The screen-type debris breaker should be shifted in accordance with the shift of fan segment. The iron bars can, of course, be used again and again.

Discussion

For developing countries which are suffering from flood disasters it is suggested that:

(1) the appropriate format, technology and procedure for damage assessment and failure analysis be developed and employed;
(2) appropriate systems for land management data and information be developed and employed, taking into account social, economic, and cultural aspects and future development of the flood prone regions;
(3) appropriate measures for public education and information on flood and related subjects be developed and employed to promote incentives for flood disaster prevention and efficient and impartial use of water resources;
(4) appropriate institutions to promote a sense of responsibility of the people in charge of flood prevention and water and river basin management;
(5) hydrologic and hydraulic monitoring systems and networks be developed and expanded. In order to encourage the people in charge of monitoring services appropriate institutions be established;
(6) international co-operation in developing and managing international rivers such as Ganga-Brahmaputra river system be developed and expanded;
(7) an international team specialized in river basin management and flood disaster prevention be organized and despatched to flood disaster stricken countries immediately after disaster, in order to provide technical and institutional assistance to prevent any recurrence of similar disasters;
(8) low-cost structural measures be developed and employed.

REFERENCES

AHMAD, M. (1985) *The coastal tragedy*. Community Development Library, Dhaka

ASHIDA, K., TAKAHASHI, T. and SAWADA, T. (1981) Process of sediment transport in mountain stream channels. IAHS-AISH Publication no. 132, 169 pp

ESCAP/WMO, TYPHOON COMMITTEE (1986) Report of the Extraordinary Session, Manila, Philippines

MINISTRY OF PUBLIC WORKS AND POWER (1969) Mt Kelut Volcanic Debris Control Project, Feasibility Study Report. Indonesia, 18 pp

GOVERNMENT OF WEST BENGAL (1981) Flood forecasting and warning; the social value and use of information in West Bengal

MATHUR, S.K. (1981) UNDRO News, May/June 1985

MIZUYAMA, T., ISHIKAWA, Y. and FUKUMOTO, A. (1988) Estimation of mudflow hazard areas associated with volcanic eruptions and mitigation of the hazards. PWRI no. 2601, Japan, 26 pp

RANGANATHAN, V. (1985) Impact of tropical cyclones on Maharashtra State, Seminar on a systems approach to tropical cyclone preparedness problems. National Institute for Training in Industrial Engineering, India, 21 pp

SIKANDER, A.S. (1983) *Floods and families in Pakistan – a survey.* Disasters, Foxcombe Publications, vol. 7, no. 2, 101 pp

SINHA, A.K. and AVRANI, S.U. (1984) *The disaster warning process: a study of the 1981 Gujarat cyclone.* Disaster, Foxcombe Publications, vol. 8, no. 1, 73 pp

UNDRO (1985) UNDRO News, May/June 1985

UNDRO (1987) Infrastructure damage and rehabilitation, Bangladesh flood 1987

WATANABE, M. (1987) Some problems in disaster prevention and water resources development in the river basins in which much sediment yield and transportation take place in developing countries. Shin sabo vol. 39, no. 4, Japan, 34 pp

WMO/ESCAP, PANEL ON TROPICAL CYCLONES (1986) Report of the Thirteenth Session, Rangoon, Burma

YAZAWA, A., MIZUYAMA, T. and KITAHARA, I. (1986) Study on evaluation of effects of countermeasures for debris flow by the simulation method of debris flow. PWRI no. 2392, Japan, 18, 15 pp

The Hydrology of Disastrous Floods in Asia – An Overview
—

NAGINDER S. SEHMI

Hydrology and Water Resources Department,
World Meteorological Organization, Geneva, Switzerland

Introduction

THE HYDROLOGY of disastrous floods in the tropical cyclone and monsoon countries of Asia is changing rapidly, mainly as a consequence of the high rate of economic development accompanied by growing urbanization. It is becoming increasingly 'intensified' and 'localized' both in terms of cause and effect. For thousands of years the problem of flood disasters has been associated with the plains in the middle and lower reaches of the major rivers which now accommodate approximately half the population of the world. These people have known their rivers for centuries and have, in most cases, tamed them to an extent that the major rivers no longer appear to cause direct major disasters. Feelings of security from floods have increased population pressure beyond any expectations. This has led to the development of land and water resources beyond what is desirable, at the expense of shrinking forests, eroded pastures and polluted rivers and lakes. Traditionally, operational hydrological activities are geared to floods in large rivers, whereas the major flood disasters now occur in densely populated areas in small river catchments. A Persian proverb aptly describes the problem: 'In the ants' house the dew is a flood'.

The Problem

The damage caused by floods in Asian countries was estimated at more than US $5 billion in 1981, and is steadily rising. At the same time the areal extent of flood disasters is increasing rapidly. In India, for example, the total area subject to flooding doubled from 20 million ha in 1971 to over 40 million ha in

1981 (Darryl, 1985). This has happened as the number of large dams (over 15 m) alone has risen from about 1700 to 1950 to over 23 200 in 1982, not to mention the more numerous, smaller than 15 m high dams. Then there are the flood control levees whose total length exceeds 200 000 km (ESCAP, 1986). Failures of these structures, which is not uncommon, has produced catastrophic floods.

The countries in Asia are experiencing a reduction in the safety of existing flood control structures, and at the same time an increase in the risk of disasters from flash floods in small catchments and densely populated areas, in particular in the cities. In some cases the flood control structures have contributed to the magnitude of flash floods, especially when they impede runoff drainage. Many population centres in Asia are finding it more risky to be totally dependent on levée and canal systems for flood protection. This is because most flood protection works were built to standards applicable to pre-development, when the basin possessed undisturbed hydrological conditions. But now these standards – the design criteria – are being exceeded frequently. Intense rainfall on a small steep catchment upstream of a poorly drained urban centre causing a devastating flash flood is becoming a common occurrence. Such flooding is the cause of about 90 per cent of lives lost through drowning in tropical cyclone countries.

Major changes in the use of water and land resources have adversely affected the environment, in particular the hydrological regimes of many basins. These adverse trends, in turn, have aggravated the flood problem, as shown by Uehara (1987) in Table I.

For example, in the Mekong Basin, a large portion of the forest cover has been removed over the last 30 years as timber to create agricultural land. The negative hydrological consequences of deforestation and incorrect land-use policies have resulted in an increase in the sediment load of rivers, siltation of irrigation works and reservoirs, and an increase in the flash flood hazard (Mekong, 1987). The metamorphosis of hydrology and the resulting problems are addressed according to the size of the river basins.

Major rivers

All countries in this region experience a monsoon climate with great seasonal rainfall variations. Between 70 and 90 per cent of the total annual rainfall occurs during the wet season (May to October). As would be expected, river flows show even more pronounced seasonal variations which amplify difficulties in the control of floods.

The large amount of effort and investment injected into flood control measures in the major river basins of the region are well known. For example, the vast low-lying alluvial plains of China drained by seven major river systems (the Yangtze, the Yellow, the Huai, the Hai, the Laio, the Pearl and the Songhua Rivers) cover an area of over one million square kilometres, and

TABLE 1 *Increasing rates of population and assets in flood risk areas in Japan*
(Percentage of whole country)

	1960	1970	1980	1985
Risky area	10	10	10	10
Population	44.7	46.3	48.2	48.7
Assets	51	63	72	75

contain half the population of China and most of its important cities. The ground levels of the plains are generally lower than the flood stages of the rivers. Consequently, these areas have been safeguarded by dykes with a total length of over 160,000 km, most of them built since 1949. In addition, numerous flood water storage and diversion facilities have been created to contain the overflows. In short, flood control structural measures have made the vast plains largely safe from floods of a 10- to 20-year return period (Qian Zhengying, 1983).

The dyking has triggered aggradation of river beds, thus reducing the safety of the existing structures. It is estimated that, if the rainstorm of 1981 (China, 1987) had extended to cover the middle and lower reaches of the Yangtze, as occurred in 1954 (a 40-year flood), the probable losses would have been in the tens of billions of dollars; more than half a million people would have had to have been evacuated, and only the safety of the Jingjiang Dyke and Wuhan city could have been ensured. In addition, river channel obstructions caused by vegetation growth and deposition of debris (naturally or by man) have become paramount in amplifying disasters even from a flood of low return period.

Cognizant of such possibilities, in particular of the decreasing safety of existing flood control structures, China has developed an elaborate and modern, but very complex, flood forecasting service for monitoring the entire country on a real-time basis. In fact, the hydrological forecasting and warning systems are seen as the 'eyes and ears' of the country, without which it would be difficult, if not impossible, to mitigate a flood disaster (China, 1987).

Small rivers

Monsoon rains, tropical storms and typhoons usually result in very intense rainfall in a short time over a small area, leaving little time for the collection of data and issuing of forecasts. Hence the forecast or warning lead time is frequently non-existent or very short indeed. Such meteorological situations lead to treacherous 'flash' floods, defined as floods of short duration with a relatively high peak discharge where the time interval between the observable causative event and the flood is less than four to six hours (WMO, 1987). In some countries the flood forecasting service, with the collaboration of the meteorological services, monitor the behaviour of tropical cyclones and storms and provide generalized heavy rainfall, flash flood or storm surge warnings without delaying to prepare refined, site-specific forecasts. Much remains to be done, however, in order to increase the lead-time.

Vulnerability to flood damage in small catchments has increased because development and land-use practices have depleted their rainfall retention and runoff retardation characteristics, causing not only higher flood peaks but also shorter times of concentration. The drainage and river channel improvement works have had a similar effect on the flow. Inefficient or poorly maintained drainage systems exacerbate the situation.

Kuala Lumpur case study

The city of Kuala Lumpur (Malaysia) is prone to severe flash floods. It is situated at the confluence of three short (30 to 50 km long) and relatively steep streams which drain the small (427 km^2) catchment of the Kelang river (Figure 1). The fan-shaped catchment has the almost perfect hydrological characteristics for producing severe floods. The average annual population growth rate in recent years has been 4.3%, of which natural increase is 2.2%. The population is expected to increase from 1 million in 1980 to 1.5 million in 1990. From the earliest days Kuala Lumpur has been subject to flooding, but the extent of damage caused in recent years is beyond any expectations. The tropical storm of 1–5 January 1971 produced a record flood of 569 m^3/s over a period of 16 hours. The flood damage was estimated at over US $14 million.

Of the three reservoirs planned on the tributaries, the one on the Kelang has been completed. In addition, considerable channel improvement works have been undertaken. On completion of the project, the city is expected to be protected from a 100-year flood. But the project, involving an investment of over US $92 million, will require 13 years to complete. In the meantime, the flood damage potential is on the rise. Large-scale development for housing and industrial estates is rapidly replacing the thick vegetation of the hillsides. Figure 1 also shows the urban expansion since 1962.

These developments, both urbanization and structural measures for flood control, have evidently changed the hydrology of the catchment. Based on the flow records at Market Street and Sulaiman Bridge in Kuala Lumpur, the historical variation of the runoff pattern is represented by a sharpness index which is the ratio of instantaneous peak discharge to the daily mean discharge. Figure 2 shows that the sharpness index has been increasing since 1960 (TCS, 1987).

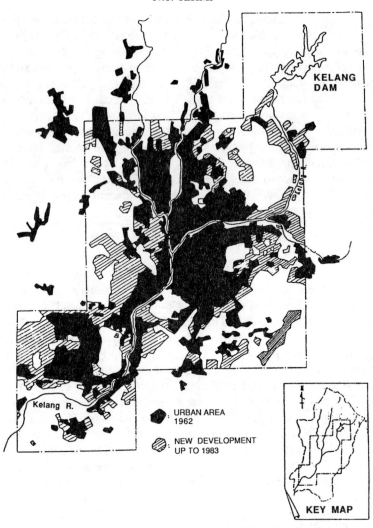

FIGURE 1 *Urbanization of the Kelang river basin, Kuala Lumpur (Malaysia)*

Flash Flood Incidents in Selected Countries

The actual extent and gravity of disasters caused by flash floods in densely populated areas other than cities is difficult to determine until countries carry out systematic surveys. A breakdown of deaths caused by major forms of natural disasters in Japan is shown in Figure 3 (NRCDP, 1983). Floods are responsible, on average, for over 60% of them. Reports on some recent flash floods attest to their seriousness.

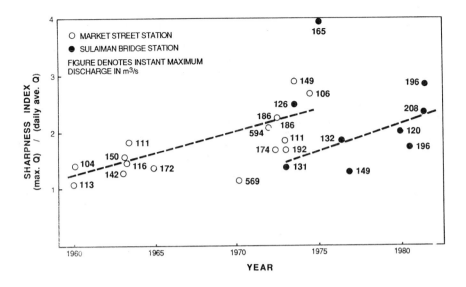

FIGURE 2 *Historical variation of flood flow pattern*

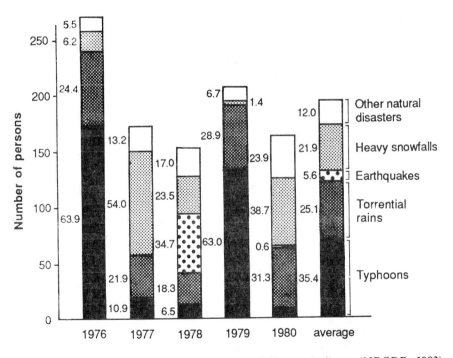

FIGURE 3 *Deaths caused by major forms of natural disasters in Japan (NRCDP, 1983)*

Intense precipitation usually results from tropical storms which frequently penetrate into the temperate regions of China. Typhoons from the China Sea give rise to extraordinary rainfalls (for example, 1605 mm in 72 hours at Linzhuang in Hanan Province in August 1975) resulting in severe flash flooding. On 11 July 1986 typhoon PEGGY shed between 500 and 1000 mm of rain in the Fujian provinces; Mei Xian was under 2 to 4 m of water for 40 hours. In 1987, floods and typhoons caused 2300 deaths, and destroyed 650000 homes. The losses were estimated at US $2400 million.

In Malaysia, other urbanized areas such as Penang, Kota Baharu, Ipoh and Johor Baharu experience increasingly severe flash floods. In November 1986, the heaviest rainfall for 15 years caused severe flash floods in the lower Trengganu and Kelantan rivers. Fourteen people drowned and 20000 had to be evacuated; the cost of the damage was estimated to be more than US $11.4 million.

All parts of the Republic of Korea experience floods which pose a severe hazard to the population and exert a major negative effect on the economic growth of the threatened areas. Floods have occurred in the Han river on an average of about twice a year, and severe flooding is experienced on an average of once every four years. The disastrous effects of floods have been mitigated by constructing major flood control works, including a system of seven dams and river training works supported by a flood forecasting service (Figure 4). However, the problem of localized flash floods in small river basins, particularly in coastal areas, is still to receive attention. In 1987, in the Chungchung and Seoul Kyunggi areas three heavy rainfalls and associated flash floods caused 243 deaths through drowning.

Flash floods, usually associated with typhoons, often take a heavy toll of life and cause much property damage in the Philippines. On 3 October 1981 for example, a flash flood in a volcanic catchment carried soil with it into the Amacon Creek causing overtopping of an earth dam. In 30 minutes, the deluge composed of mud, debris, logs and boulders destroyed houses and a market centre leaving behind 124 dead, 12 missing and the 204 injured employees of a mining corporation. A total of 622 families (3732 persons) were affected by the disaster. In September-October 1985, intense flooding in 18 townships caused an even bigger disaster. In major urban centres, particularly Manila, flash floods have become more frequent, causing serious damage. It is estimated that in Manila about 70% of the urban drains get clogged. It is feared that, should a major flood occur, it could cause a disaster far greater than anything experienced in recent times.

Tropical storms and cyclones frequently cause destructive flash floods in Thailand, which are becoming increasingly disastrous with the rapid rate of development, deforestation and urbanization. Metropolitan Bangkok suffers from flood damage about once every two years. Flooding is caused by local flow being backed up by prolonged high water levels in the Chao Phraya River. On 8–9 May 1986, the 1000-year rainstorms over Bangkok resulted in

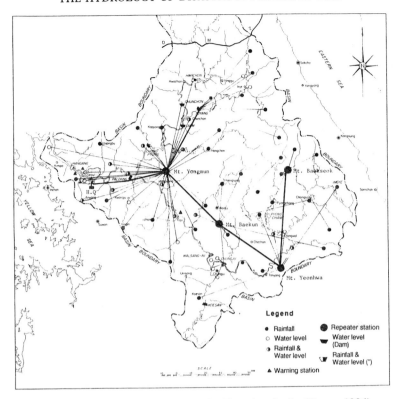

FIGURE 4 *Telemetry network in the Han river basin (Korea, 1986)*

historic floods. Two tropical depressions (BETTY and CARY) between 15 and 27 August 1987 produced severe flash floods in 39 provinces in northern Thailand and five provinces in the south resulting in widespread destruction. Table 2 shows the gravity of the situation.

Flash floods hit Afghanistan in June 1988, killing many people and 117,000 cattle, damaging 1,300 km of irrigation works in addition to roads, bridges and culverts, causing damage estimated at US$ 260 million.

Land-use Studies and Applied Research

A number of countries have undertaken *ad hoc* studies on the effect of land use on the hydrological regimes of different basins, but systematic and intensive studies to investigate the effects of changing land use on flood characteristics have been undertaken only in Japan. One such study of the flood hydrology of the Shakujii river basin (48 km^2) during its 20 years of urbanization (1958–1977) has shown that the 1977 peak discharge had

TABLE 2 *Summary report on flash floods in northern Thailand*

No.	Duration August 1987	Disaster areas			Damages							Financial mitigation by DPW* (bahts)	Notes
		Province	Amohoe	Kring Amohoe	Families	Persons	Deaths	Houses	Other structures	Agricultural area (ha)	Pets		
1.	16–27	Chieng Mai	16	1	2,099	10,140	6	151	204	33,242	2,847	248,000	four injured persons, 2621 flooded houses
2.	15–27	Phatchabon	1	1	—	—	—	—	1	12	—	—	—
3.	16–17	Lumchun	3	—	473	2,365	1	—	36	10,907	—	27,040	—
4.	23	Nan	8	3	300	1,500	5	4	2	—	109	25,000	—
5.	17–24	Phrae	4	—	—	—	2	—	13	1,950	—	4,000	—
6.	22–26	Suchotnai	3	—	5,694	28,470	—	—	8	26,300	—	15,600	—
7.	24–25	Pnayao	4	—	—	346	1	—	12	4,320	—	3,985	—
8.	22	Chiang Rai	2	—	89	—	—	—	—	—	—	104,841	89 flooded houses
9.	23–26	Tak	3	1	—	—	1	29	16	996	100	28,364	—
10.	23	Uttaraidit	1	—	—	—	—	—	—	—	—	2,065	—
Total			45	6	8,645	42,821	16	184	292	77,727	3,056	458,895	

*Department of Public Works.
1 US$ = 25 bahts.

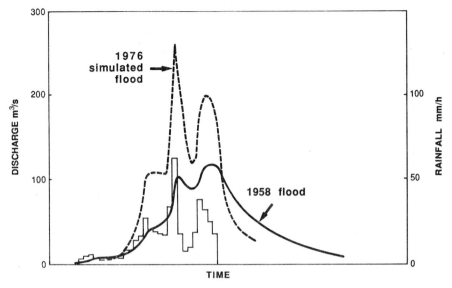

FIGURE 5 *Flood flow changes due to urbanization, Shakujii river, Japan (Uehara, 1987)*

increased by up to 2.5–3.0 times that in 1958 (Figure 5). This increase corresponds to an increase of residential area from 33% to 52% of the catchment (Uehara, 1987). Results of a similar study of the Thurumi river basin (235 km^2) are shown in Figure 6 (Kinoshita *et al.*, 1986).

Research activities related to flood forecasting in the tropical cyclone countries range from a very advanced level in Japan to only a very basic level in some other countries. In densely populated and built-up areas, the approach to flood forecasting has been considerably intensified in time and space. For example, from the results of studies of physiographic characteristics, flood-risk mapping and drainage systems in Tokyo, flood impacts are predicted using assumed values of rainfall intensity such as those shown in Figure 7 for Hong Kong. Japanese meteorologists are developing systems for forecasting rainfall for a 1 km × 1 km mesh at about five- to 10-minute intervals. From this information the hydrologists should be able to issue a more accurate and precise forecast of floods (Kinosita, 1987).

Other countries are making special efforts to develop flood forecasting and warning systems. For example, Tingsanchali (1987) has reported on the experience of calibrating and applying the SSARR model for flood forecasting on the Pong, Chi and Mun river basins in Thailand since 1981. The forecast of rainfall remains the most critical factor in the accuracy of these flood forecasts.

A significant impetus to research in flash flood forecasting was given during the Typhoon Operational Experiment (TOPEX). The hydrological component of TOPEX was implemented on the basis of three water-years starting in March 1981. The objective was to reduce the risk of loss of life and damage

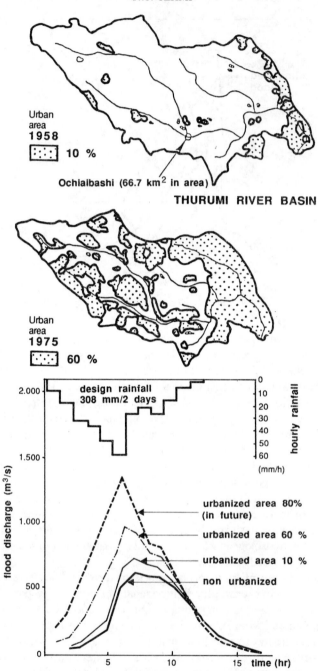

FIGURE 6 *Effect of urbanization on flood flows at Ochiaibashi (Japan) (Kinosita et al.,* 1986)

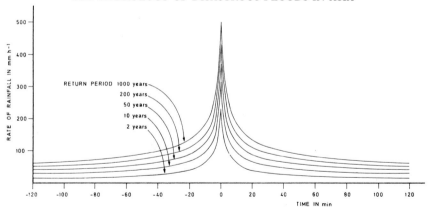

FIGURE 7 *Storm profiles for engineering applications in Hong Kong (Royal Observatory, 1986)*

from typhoon-related floods by improving the forecasting and warning capabilities of countries. Six countries designated nine river basins whose flood forecasting systems were evaluated during the three years (WMO, 1985).

In 1987 the Typhoon Committee carried out a survey of the hydrological research activities of Asian countries, and the results were reported by Sea (1987). In China, a number of flood forecasting models have been applied, tested and adapted for local use. A combination of the Xinanjiang model and the Constrained Linear System (CLS) has resulted in a Synthesized Constrained Linear System (SLCS) which is now being used operationally (Wang et al., 1987).

Development of Flood Forecasting Systems

Flood forecasting is the real-time estimation of stage, discharge, time of occurrence and duration of flooding, especially of peak discharge, at a specific point, resulting from precipitation and/or snowmelt (WMO, 1974). In Asia, as in many other parts of the world, floods are not only a curse but also at times a blessing. Flood forecasting combined with flood-plain management can reduce the curse while retaining the blessing.

Until recently, flood forecasting in tropical cyclone- and typhoon-affected countries was considered to be less important than flood preventive structural measures. All countries recognize that while structures try to 'keep the water away from people', it is almost impossible to 'keep people away from water'. In this connection, Sugawara (1978) has the view that flood control reservoirs are effective for small- and medium-sized floods but are of little value for the control of large, very infrequent events, especially when the populations they protect have attained full confidence in the security provided by the control. It

117

TABLE 3 *Major flooded forecasting projects implemented by WMO*

Country	Phase	Duration	Rivers	Assistance in US$
Bangladesh		1980–86	—	3,040,000
Burma	I	1979–84	Irrawaddy (upper)	330,000
	II	1986–89	Irrawaddy (lower)	1,056,000
China		1981–85	Yellow (lower)	700,000
		1983–86	Yangtze (middle)	900,000
India	I	1980–85	Yamuna	1,255,000
	II	1985–88	Yamuna	261,000
Indonesia		1987–90	—	998,000
Nepal		1982–87	—	1,031,000
Pakistan	II	1985–88	Indus	595,000

seems that flood disaster prevention measures, including flood forecasting, have this unfortunate characteristic: they increase the damage from large disasters. This is so because the increase of private and public properties in the flood-prone areas not only raises the damage potential but also increases the difficulty in providing flood control facilities.

While considerable effort and resources have been expended in Asia to establish flood forecasting services in major river basins, only a start has been made in providing effective flash flood forecasts. Table III shows the major technical assistance projects implemented by WMO since 1980 to set up flood forecasting services.

The total technical assistance so far given represents but a small fraction of the total national effort. Malaysia, for example, has a well organized flood forecasting service covering all the main rivers which is being extended gradually on a selective basis. Since 1971, four peninsula rivers have been equipped with mainly locally manufactured automatic telemetry and transmission systems. Flood forecasts are prepared by the Hydrological Forecast Unit using the Sacramento and Tank models. For other rivers, forecasts are made by a stage-correlation procedure. On a few rivers, dams which were not constructed specifically to control floods do help in mitigating flood disasters. However, no other major river control structures are planned for the near future. It is planned to upgrade the flood forecasting services for the entire country. Flood Warning Boards have been established at the local level to issue forecasts and warnings in each dangerous river. In addition, basin-wide hydrological studies have been initiated. In short, Malaysia has already acquired considerable experience in flood forecasting and warning procedures and has a sound infrastructure, as shown in Figure 8.

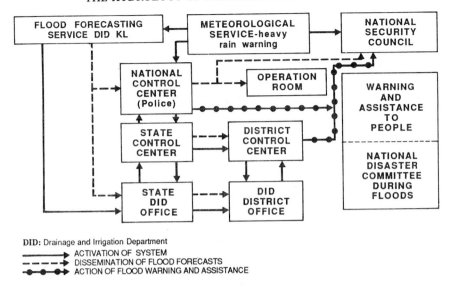

FIGURE 8 *Flood warning and forecasting information service system (Malaysia)*

In neighbouring Thailand only a few flood forecasting efforts have been made independently by various national agencies. Recent severe flash floods have sharpened the awareness of the authorities concerned of the need for provision of an efficient nationwide flood forecasting and warning service. There already exist considerable facilities and equipment which need to be fully exploited. For example, in addition to a dense network of meteorological and hydrological observing stations, the entire country is monitored by a system of eight weather radars (Figure 9). There is an immediate need for national planning and development authorities to evaluate the hydrological and meteorological facilities, the experience and the expertise available in the country, in order to mobilize the resources required to mitigate flood disasters.

In China, the Republic of Korea, Hong Kong and Pakistan the flood forecasting services are rapidly being modernized, while Burma, India, Indonesia and the Philippines have made a good start. In general, the quality of water, which forms a major component of flood disasters, has not yet aroused concern in most of the countries.

Concluding Remarks

Progress towards the orientation of hydrological activities to effectively mitigate the disastrous effects of flash floods in Asia does not seem to correspond to the rapid increase of flood incidents. More funding and greater

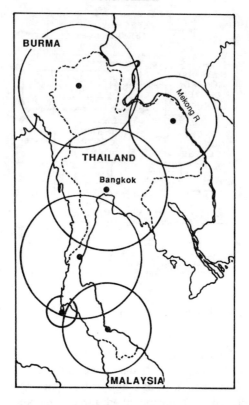

FIGURE 9 *Weather radar coverage of Thailand*

emphasis is required for the establishment of reliable and rapid (computer) systems for data processing and analysis and for the application of forecasts of meteorological conditions, especially the tracks, locations and coverage of rainstorms and tropical cyclones (typhoons). Some countries have recognized and adopted the approach that an efficient flash flood forecasting service should be geared to flood disaster prevention systems in urban areas, as it is becoming increasingly difficult without them to manage and operate the water supply and drainage systems, not to mention pollution control.

The primary task is to identify all densely populated locations which are subject to flash floods, to study their hydrology, to delineate the flood-prone areas and to prepare plans with priorities for the installation of custom-tailored flood forecast/warning systems in each location.

Based on the experience already acquired, flood forecasting services need to be extended to all flood-prone river basins. To do so it is necessary to integrate the meteorological and hydrological data being produced by various agencies in the country, in order to improve the lead time of flood forecasts.

Opportunity should be taken to apply the modern technology needed for upgrading synoptic meteorological and hydrological networks. Based on such techniques as radar-rainfall measurement systems and high resolution satellite sensing, which can provide invaluable information on storm movement, coverage and rainfall intensities even at a relatively small (metropolitan) scale, application of quantitative precipitation forecasts (QPFs) should be promoted. At the same time, more sophisticated hydrological models which will accommodate QPFs should be introduced.

There is also the need to develop the framework and effective interagency-based co-ordination which will permit flood forecasts and warnings to be disseminated to and received by all agencies concerned with flood emergency and relief operations and by the affected people. A programme of educating the public is equally important to ensure that the people concerned clearly understand the significance and meaning of a flood forecast, an advisory, a warning, an alert (WMO, 1988) and the measures they are expected to take in order to minimize loss.

As part of the various operational hydrology activities, it is becoming increasingly important to undertake studies of the effects of deforestation, urbanization and changing land-use on the hydrology of the area concerned, and thus on the intensity and duration of floods.

Finally, international organizations are orientating their programmes in operational hydrology according to the changing needs of countries. In the past the emphasis has been mainly on promoting international co-operation and strengthening the capabilities of the national Hydrological Agencies in a relatively broad sense. Naturally, the guidance material generated by them pertains to large hydrological phenomena. They are now focusing on small-scale facets of hydrology. The operational technology normally applied to flood forecasting in large river basins is not suitable for forecasting flash floods on small basins. This applies equally to the measurement and monitoring the flows in urban areas. In short, it is the operational aspects of the changes in hydrology that the programmes are now aiming at.

Acknowledgements

The author acknowledges with appreciation the kind authorization of the Secretary-General of WMO to submit this paper. The paper is based mainly on the author's mission to six typhoon-affected countries in November/December 1987. The author therefore wishes to express his gratitude to those countries which have given valuable information to ESCAP, which organized the mission, and to other members of the mission for the material they have provided.

REFERENCES

CHINA (1987) *Brief introduction to hydrological information and forecasting in China*. Ministry of Water Resources and Electric Power, Beijing, China

DARRYL, D'MORTE (1985) *Temples or tombs*. New Delhi, p. 14

ESCAP (1986) Environmental issues of water resources development in ESCAP region. *Document* E/ESCAP/NR. 13/11, Bangkok, Thailand

KINOSITA, T. *et al*. (1986) *Hydrology of warm humid islands*. National Research Centre for Disaster Prevention, Japan

KINOSHITA, T. (1987) One-kilometre meteorology (personal communication)

KOREA (1985) *Flood forecasting and warning system of Han River*. Han River Flood Control Office, Ministry of Construction, Seoul, Republic of Korea

MEKONG (1987) *Annual report*. Interim Committee for Co-ordination of Investigations of the Lower Mekong Basin, Bangkok, Thailand

NEMEC, J. (1986) *Hydrological forecasting*. Reidel, the Netherlands

NRCDP (1983) *Outline of recent activities of National Research Centre for Disaster Prevention*. NRCDP, Japan

QIAN ZHENGYING (1983) *The problems of river control in China*. Ministry of Water Resources and Electric Power, Beijing, China

ROYAL OBSERVATORY (1986) *Meteorological research, publications and consultant services*. Pamphlet Serial No. 2-8, Hong Kong

SEA, C.H. (1987) *Report on research under hydrological component*. ESCAP/ WMO Typhoon Committee, Manila, Philippines

SUGAWARA, M. (1978) On natural disaster – some thoughts of a Japanese. *WMO Bulletin*, April 1978, Geneva, Switzerland

TCS (1987) *Flood risk analysis and mapping in some countries*. Mission report, ESCAP/WMO Typhoon Committee Secretariat, Manila, Philippines

TINGSANCHALI, T. (1987) Development of flood forecasting and warning systems in Chi and Mun river basins – Northeast Thailand. *Water Res J* ST/ESCAP/SER. C/154 (Sept. 1987), pp. 31–44, UN, Bangkok

UEHARA, S. (1987) Prevention of disasters related to rivers. *Technology for disaster prevention*, Volume 11, (Sept. 1987), p. 81, NRCDP, Japan

WANG JUEMOU *et al*. (1987) *The synthesized constrained linear system (SCLS)*. Ministry of Water Resources and Electric Power, Beijing, China

WMO (1974) *International glossary of hydrology*. WMO-No. 385, Geneva, Switzerland

WMO (1981) *Flash flood forecasting*. (By Hall, A.), Operational Hydrology Report No. 18, WMO-No. 577, Geneva, Switzerland

WMO (1985) *Activities under the hydrological component of TOPEX*. Report No. TCP-20, WMO/TD-No. 37, Geneva, Switzerland

WMO (1988) *Technical Regulations*. Volume III. WMO-No. 49, Geneva, Switzerland

Conceptual River Routing Model 'CWCFF'

R. RANGACHARI, R.S. PRASAD,
T.K. MUKHOPADHYAY
and I. RISHIRAJ

Central Water Commission, New Delhi, India

Introduction

HYDROLOGICAL forecasting is an important outcome of applied hydrology. Such forecasts are needed in relation to flood warnings, regulation of runoffs, operation of storage systems for multipurpose uses, water supplies and other such forms of water management. Hydrological forecasting thus forms the most essential input in the operation and management of water resource systems.

With the advent of fast computing aids like electronic computers many conceptual models have come into vogue, each claiming to have one or more advantages: reliability, accuracy, simplicity or timeliness. Research is continuing in many reputed institutions and the number of operational models is increasing, while existing models are being improved.

Water resources systems – either in their natural forms or with such man-made changes as come about – are complex and perhaps no two systems can be said to be identical in their behaviour. For the same reason no two conceptual models can be expected to give identical hydrological forecasts. The ultimate test for their applicability for any particular case or situation will be the 'hindcast' verification for the system itself.

Any natural or man-made system can be simulated by proper identification of the physical and interactive processes, inputs/outputs and the related sub-systems. Physical processes can be to a considerable extent replicated using procedures with parameters. Such replicated procedures of the system can be utilized for extension of the future behaviour of the system, when the inputs for a particular period are correctly known. When this can be done accurately, the process for future behaviour, i.e. forecasting, can become a reliable tool for proper water management.

Mankind has long been engaged in the use of mathematical models for forecasting river flow: for flood control, reservoir regulation and efficient water resources management. At least 20 different types of models are in use at present. Although these models at a first, superficial, glance may look very different, they basically function along the same principles, having as their major sub-systems rainfall-runoff, river reach routing and reservoir regulation.

An automatic telemetered data collection system has been set up in the Yamuna river, upstream of Delhi, the capital of India. The Yamuna river has a drainage area of 19,300 km^2 up to Delhi. Delhi often experiences floods. In 1978 a severe flood swept through Delhi causing extensive devastation. This river has a steep slope of 1 in 16 in the first 25 km, which gradually flattens out to 1 in 500 at the foothills; then the river flows through flat terrain to Delhi, having an average slope of 1 in 4,000. Average rainfall for the whole year is 117 cm and average rainfall during the south-west monsoon is 75 cm. In this Delhi (Yamuna) flood forecast system conceptual models already in use were first considered.

The better known models like Streamflow Synthesis and Reservoir Regulation Model (Developed by US Army Corps of Engineers, Portland, Oregon, USA), HEC-1F Model (Developed by Hydraulic Engineering Centre, US Corps of Engineers, Davis, USA), NAM-S11F Model (Developed by Danish Hydraulic Institute, Denmark) and Non Linear Catchment and Routing Model (Developed in Czechoslovakia) were even adapted and calibrated for real time use in Yamuna context and tested as to their reliability for the river and for the case in question. Fast computations were facilitated by use of HP-1000 F series mini computer installed and in use. While using the well known models, Indian engineers faced difficulties due to availability of only limited physical and hydrometeorological data. Moreover, these models gave one difficulty or other, as the catchment and river characteristics under which these models had been developed were not exactly the same as those of the Yamuna. This necessitated the search for an Indian conceptual model and encouraged Indian engineers to develop a new but simpler technique suitable for the Indian field conditions of the Yamuna and Delhi.

Model Formulation

The fundamental concept of the CWCFF1 (Central Water Commission Flood Forecast 1) model is based on the fact that fluid tends to maintain uniform levels under free conditions. When fluids of two different chambers with different levels are allowed to mix freely, their new levels, after a certain time interval, will depend on their levels at the beginning of the time interval and the time lapse.

In this model a river channel is visualized as comprising of a number of parallel slices. As water flows through these slices various processes take place simultaneously between each two adjoining slices; major processes are identified as follows:

(1) Water flows from higher elevation to lower elevation;
(2) Some of the water is absorbed and retained in the dry river banks;
(3) Some of the water evaporates;
(4) Some of the water spreads away from the river bank and returns to the main stream when the river stage falls;
(5) Some of the spilled water crosses the river bank and is held up in pits and low-lying areas, never returning to the main stream;
(6) At some places water takes a separate course and joins the river downstream.

The above processes are dependent on the following physical processes:

(a) process 1 depends on the difference of levels of water in the two adjoining slices and the slope of the river bed;
(b) process 2 depends on the dryness of river banks, which in turn depends on the previous maximum water level and the lapse of time since its occurrence;
(c) process 3 depends on the dryness of the air and the air temperature;
(d) processes 4, 5 and 6 depend on the slope of the river, bank conditions and river characteristics.

In this model, processes 1, 4, 5 and 6 are combined together as characteristics of a river channel at a particular location which remain almost unchanged.

Slicing of the river channel is shown in Figure 1: slice $A\,A_1\,B_1B$, slice $B\,B_1\,C_1C$, slice $C\,C_1\,D_1D$, etc. Two consecutive slices $A\,A_1\,B_1B$ and $B\,B_1\,C_1C$ are taken out and their end elevation is shown in Figure 2 with river bed horizontal. The vertical planes, $A\text{-}A_1$, $B\text{-}B_1$ and $C\text{-}C_1$ are assumed to be sealed at the beginning of the time period. Now the vertical plane $B\text{-}B_1$ is removed and the two reservoirs are allowed to balance during period t (Figure 2). If the initial volume of water in the two reservoirs at the beginning of the period are Q_1 and Q_2 respectively, after balancing, the volume of water in each reservoir will be equal to $(Q_1 + Q_2)/2$. Now the system of the two reservoirs are given a tilt equal to the slope of the river bed. This operation is done along with the removal of the vertical and continued for time period t. A volume of water equal to $(Q_1 + Q_2)/2 \times Rn$ will flow from slice 1 to slice 2 (here Rn is a slope factor). At the end of time period t, the vertical partition $B\text{-}B_1$ is put back. This balancing process continues in series starting from the beginning of the river channel to the end of the river channel.

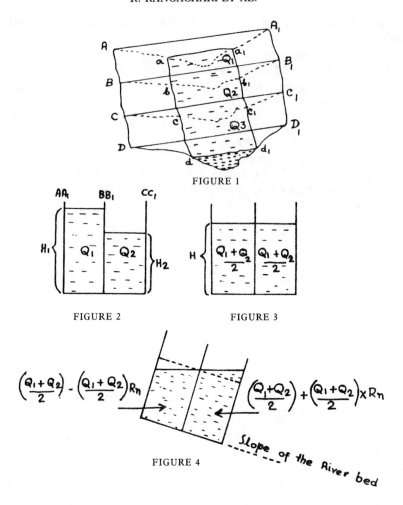

FIGURE 1

FIGURE 2

FIGURE 3

FIGURE 4

In the procedure explained above for a fixed time period t the new volumes of water in each chamber can be computed and can be continued for the entire period of computation T. The volume of water passing from one slice to the other is a function of the average of the initial volumes in the two slices and a slope factor Rn, which represents the river characteristics.

For process 2, records of previous maximum volume for each slice are stored on record and whenever the new volume exceeds the previous volume, the loss of water due to absorption in dry banks is obtained as under:

$$Vf = Vi - (Vi - B) \times Zn$$

Where Vf is the volume after absorption;
$\quad Vi$ is the initial volume;
$\quad B$ is the previous maximum recorded volume; and

Zn is the factor for water absorption. This will be executed during rising trend only.

When *Vi* does not exceed *B*, the volume of *B* is depleted by an amount equal to $B \times Ef$, where *Ef* is an evaporation factor.

Parameters Involved

Parameters involved in the CWCFF1 model are the following:

Rn,   a routing coefficient, which indicates what fraction of the average volumes of the two adjacent slices will be transferred from one slice to the other;

Zn,   the factor for absorption in the dry banks. It indicates what fraction of the total volume of water, during rising of water level, will be absorbed in the dry banks of the river;

Ef,   a coefficient for reduction of the water in the banks of the river during recession. This fraction of the total previous volume is decreased in every computation period;

NSTP, indicates how many computations to be done in one hour, i.e. computation period for one operation is (60/*NSTP*) minutes;

NC,   number of slices along the river channel.

Model Calibration

In the Yamuna catchment above Delhi, the CWCFF1 model is being tried for river reach routing in the reach between Kalanaur to Delhi. The length of the reach is about 196 km with a catchment area of 6,630 km^2 along the river. The model parameters are calibrated by trial and error method. To start with, the width of each slice was taken up as 1 km. The final parameters found out are stated in Table 1.

As a thin layer of water will adhere to the river channel, there will be no transfer of water from one slice to the other if the discharge limit is not exceeded for the number of steps (*NSTP*) as indicated below:

Number of steps per hour	Discharge limits
12	1200
11	1000
10	950
9	850
8	600
7	200
6	0

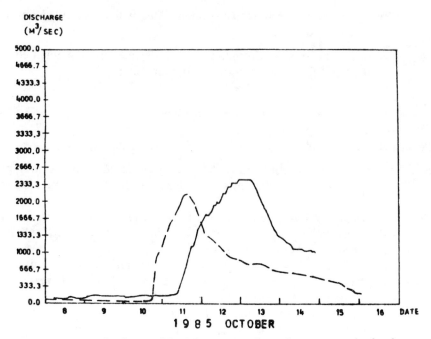

FIGURE 5 *Rn increased by 0.3.* ———, *observed;* — — —, *simulated*

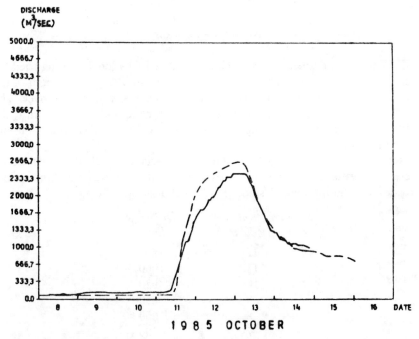

FIGURE 6 *Zn reduced to 0.001.* ———, *observed;* — — —, *simulated*

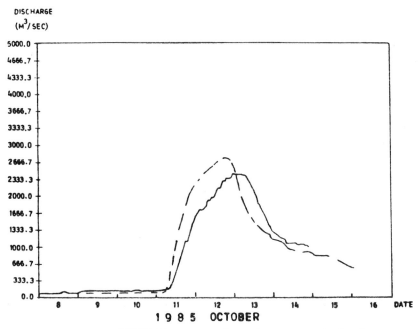

FIGURE 7 *NSTP 16.* ———, *observed;* — — —, *simulated*

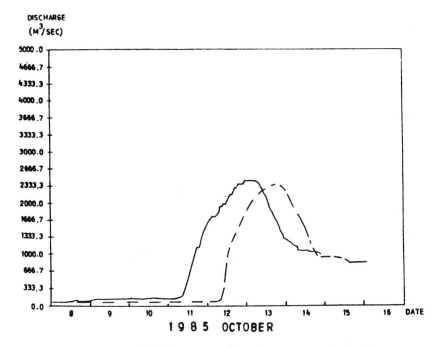

FIGURE 8 *NSTP 8.* ———, *observed;* — — —, *simulated*

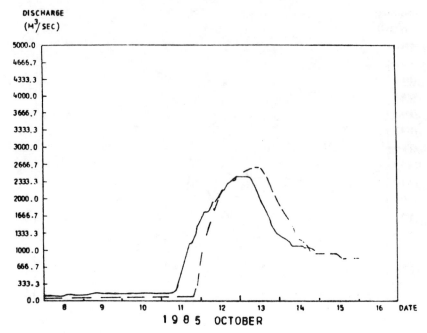

FIGURE 9   *NS 250.* ———, *observed;* — — —, *simulated*

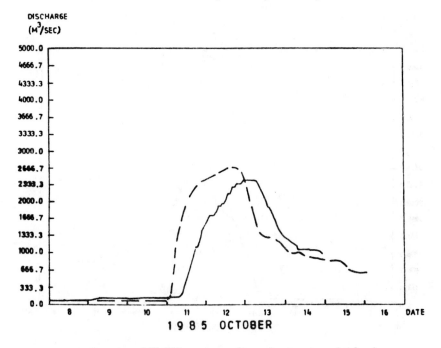

FIGURE 10   *NS 150.* ———, *observed;* — — —, *simulated*

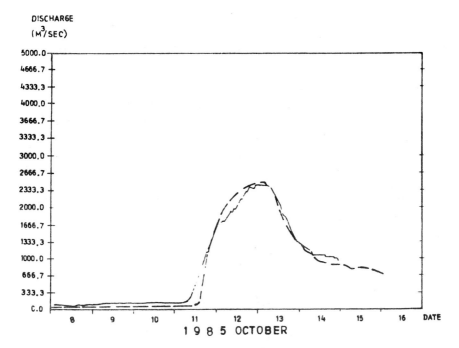

FIGURE 11 *NSTP 12.* ————, *observed;* — — —, *simulated*

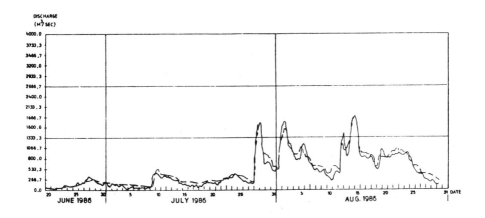

FIGURE 12 *Final parameters.* ————, *observed;* — — —, *simulated*

Effect of Parameters

While calibrating the model, the following effects of different parameters are being observed:

(1) when the number of slices is reduced, i.e. the width of the slice is increased, the simulated peak is formed earlier and the rising limb becomes steeper. When the number of slices is increased, the peak forms later with slightly reduced slopes (see Figures 9 and 10);

(2) when *NSTP* is reduced, the peak is delayed with lower value and when *NSTP* is increased the peak is formed earlier with higher magnitude (see Figures 7 and 8);

(3) when the *Zn* value is reduced, the excess of volume of water is available in a simulated peak (see Figure 6);

(4) when the *Rn* value is increased, the peak is reached faster with reduced value (see Figure 5).

(5) Figures 11 and 12 show simulation with final parameters.

Future Developments

CWCFF1 is not only a river channel routing model. The methodology can be extended to work as a complete watershed model, and is described below:

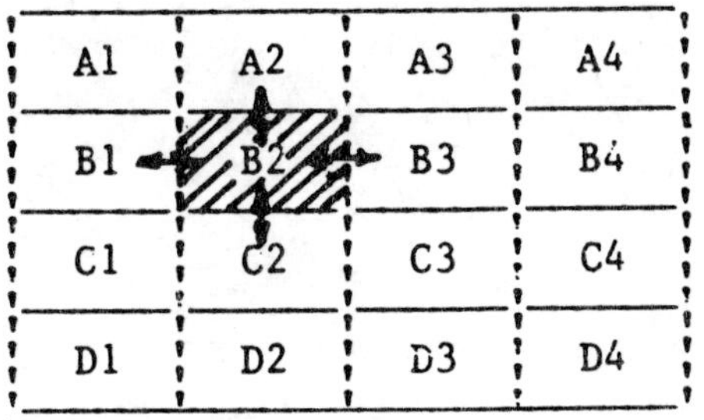

FIGURE 13 *CWCFF1 as a complete watershed model. See text*

A river catchment is to be divided into a grid of equal areas. Starting from one corner of the top of the catchment the water contents of each square compartment will have to be balanced considering the four adjacent compartments as shown in Figure 13. This procedure will have to be continued up to the end of the river channel for each computation period. While applying this model to a catchment as a complete catchment model the following procedure is to be followed:

(1) elevation of the centre of each square grid is to be determined from a topographic map;

(2) boundary conditions for each square are to be identified;

(3) proper separation of surface runoff and base flow is to be done, and both types of flow are to be routed separately up to the river channel;

(4) for square grids containing the river channel, there will be no base flows; the base flows from the adjoining square grids will end up as surface flow in the river channel;

(5) rainfall over each grid is to be computed in a simplified way by clustering a group of square grids together, based on the topography of the catchment.

Summary and Conclusions

(1) Muskingum and other simplified models are not relevant in the Upper Yamuna case because they are suitable for upstream controlled flows, whereas in the Upper Yamuna case, downstream levels actually govern the drainage.

(2) The down stream constraints are also not uniform. They vary from section to section along the course of the river and regime of the channel.

(3) It is further mentioned that the river of the type represented by Upper Yamuna can be better modelled by dividing the reach in compartments as is found in the concertina type of model situation.

(4) Since the topography remains largely unchanged for the particular volume of flow and the rates of discharge, it can simulate the physical process with fewer parameters.

(5) On the above basis, the formulation of the model for Upper Yamuna was done; the model has been field-tested and has given very good results that correspond to prototype behaviour.

(6) Most of the models work satisfactorily for simulation and forecasting of flood when there is a continuous high flow. But all the four models (as tried in the Upper Yamuna catchment) failed to simulate the river flows at the start of the flood season or when a high flood came after a long recession. CWCFF1 model has been developed so that it is capable of adjustment for the absorption losses in the river beds. It is observed that the CWCFF1 model worked satisfactorily during both low flows and high flows.

Further close monitoring during the next flood seasons will be carried out for refinement and modification of the model

Methodology for the Short-Term Warning and Estimation of the Level and Consequences of the Accidental Pollution of Water Bodies

A.M. NIKANOROV and V.A. KIMSTACH
Hydrochemical Institute, Rostov-on-Don, USSR

CHANGES in hydrological conditions, resulting from both natural and man-made causes, may be divided into three spatio-temporal groups, each of which requires a specific type of monitoring: global (background monitoring), regional (regime monitoring) or local (short-term monitoring). The aim of global monitoring is to study slow background changes. These data are the basis for the system of regime (or regional) monitoring by which are determined the seasonal, annual and interannual changes in the water quality of a given region. In turn, the changes caused by discharges of pollutants of an anthropogenic nature occur within the context of the regime changes. Discharge of anthropogenic pollutants may be both continuous (systematic) or momentary (random), such as an accidental discharge of pollutants into a water body at a concentration far exceeding permitted standards. A continuous discharge may be related to the regime monitoring, while a momentary (or accidental) discharge may be of a spontaneous or periodic nature with respect to both the time and the site of its occurrence.

Control of accidental discharges is extremely difficult, if not impossible, due to the large number of rivers, lakes and water courses exposed to man's influence.

Momentary accidental discharges, subjects of short-term or real-time monitoring, have certain peculiarities. They are difficult to predict; occur in a wide variety of circumstances; involve many different pollutants; and act for a relatively short time upon the environment, depending on their volume.

Of course, the environment cannot immediately cope with the impact of a sudden accidental discharge: the ecosystem needs time to adapt to the new conditions. Such conditions are very dangerous and often lead to the greatest damage. Hence, a pollution control system must be adapted to the real-time situation and the quicker the consequences of the accident are eliminated, the less damage will be done. In comparing the requirements of such a system with the capacities of ordinary systems of transmitting and receiving information, one may conclude that it is practically impossible to achieve this goal, if only because the time of data transmission and receipt can exceed the lifetime of the accident by a considerable margin. The detection and study of accidental discharges require a specific methodology to be developed; such a methodology must take into account the need for fast reactions to rapid changes in the state of the water ecosystem. It is necessary to include the conservation and environmental protection authorities in the organizational system in order to implement effectively any pollution mitigation methodology. According to the generally accepted definition, therefore, such a scheme cannot be classified as 'short-term monitoring'.

To limit the damage and cost-effectively liquidate the accident, it is necessary to collect data characterizing the state of the water body during the accident as well as data on the background conditions prior to the accident. Thus, the short-term control service should co-operate closely with the regional monitoring service.

Within the territory of the USSR there are about 0.5×10^6 water bodies on which a large number of water users – potential accidental polluters – depend; most water bodies may also be exposed to non-point sources of pollution such as agriculture. On this basis, the problem of establishing and maintaining short-term pollution monitoring and control systems covering all the water bodies of a given region would be enormous. On the other hand, short-lived water-pollution incidents are usually highly concentrated. Although they may cause sharp deviations in the ecosystem, the actual damage they cause is generally greatly limited in space. Such impacts occur mainly in densely populated regions, where there are normally many abstractions and discharges and therefore a need for short-term monitoring by identifying priority water bodies and setting up observational sites on them. However, an accidental pollution incident can spread in a river system to places far from the site of the incident. A short-term control system must therefore be provided to predict and monitor extremely high pollution (EHP) incidents with respect to both the hydrological factors and the toxicity of the substances entering the water body.

A continuous discharge from a certain position on a river presents a short-term monitoring system with a noise masking the useful signal – of the accidental discharge. The larger the continuous discharge at a constant rate, the lower the signal-to-noise ratio. However, the objective of a short-term control system should not be simply to monitor the absolute level of the pollution, since the concentration of the determinants concerned do not always reflect the important temporal parameters. Monitoring the sharp changes in the state of the water body is normally much more important.

The goal of short-term monitoring is thus the timely discovery of radical changes in the state of a water body in order to avoid the considerable economic and ecological damage that can result from pollution incidents. Determining the reasons for these changes and the possible consequences as well as making recommendations for short-term measures to protect the ecosystems of the water bodies are also very important.

To achieve this goal, the following problems must be overcome:

(a) Detection of the initial location and the spread of the pollution incident;
(b) Short-term assessment of the impact of the incident upon the ecosystem;
(c) Preliminary identification of the nature and level of the pollution;
(d) Preliminary prediction of the spread of the pollution (EHP);
(e) Dissemination of warnings concerning the occurrence and spread of an EHP incident;
(f) Determination of the character and source of the pollution in more detail;
(g) Determination of the spread of the pollution and prediction of its consequences;
(h) Working out recommendations for short-term measures to protect the ecosystem of the affected water body and to limit the negative impact upon the human population;
(i) Making proposals for the organization and control of pollution consequences;
(j) Improvement of the control system.

A scheme fulfilling these requirements is shown in Figure 1. When creating such an information system, the principles of optimum organization taken from other information systems should be used.

This monitoring system should consist of two subsystems. The first (peripheral vision) is comprised of the complex of methods and means reacting to the changes in the water body. Ideal from this point of view are those means that permit not only the establishment of the fact of rapid change in the ecological state, but also the estimation of the spatial parameters of this change. From this point of view the important systems for short-term monitoring are those in which remote sensing is carried out from artificial satellites (AS) or observations are made from aeroplanes or by helicopters for smaller water bodies, the latter method giving the best spatial resolution (OA-1) (Figure 2). Of course, observations may be also performed from boats and from river banks: the basic advantage of this is that the contact measurements allow problems (a) to (c) to be combined.

Remote-sensing methods can detect and outline the sites of ecocystem changes provoked by pollution. However, these techniques have not yet been perfected. Not every incident caused by extremely high pollution can be detected by remote sensing and an observation of an EHP should be made by another means. Contact observations are important; the water body should be systematically sampled at representative points. These points should be

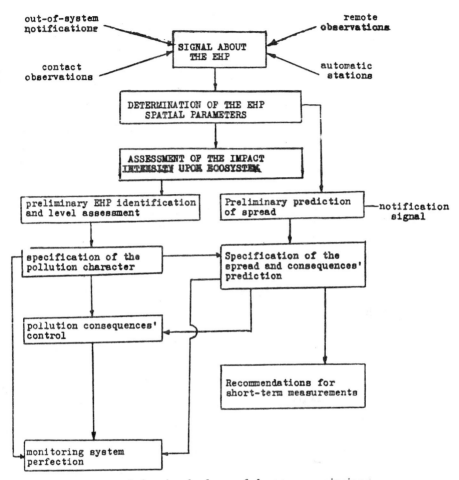

FIGURE 1 *A functional scheme of short-term monitoring*

selected on the basis of a detailed study of both water bodies and anthropogenic pollution sources with the help of mathematical modelling methods and on the basis of data from preliminary and experimental observations of the water body. The timeliness of primary data increases greatly when use is made of mobile hydrochemical and hydrobiological laboratories and of helicopters and ships.

Information about the appearance of an EHP may also be received with the help of automatic multiparameter water quality monitors. In the USSR, in particular, automatic analyzers of the AMA-201 type are installed which allow the determination of up to 17 water quality parameters automatically. Another method of automatically receiving primary information is the transmission of biological sampling. Installation of such a system at the sites of potential

137

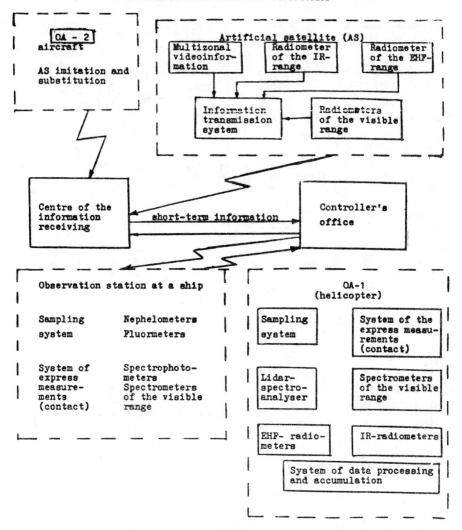

FIGURE 2 *A remote link of short-term monitoring*

accidental discharges provides for a permanent system of control.

To increase the information content of the primary measurements when detecting an EHP, mathematical methods may be developed and employed which determine relationships between different water quality parameters. In this case the primary information is divided into two types: the basic primary parameters measured experimentally and the parameters calculated from the primary ones necessary for the detection of the EHP. The number and specifications of the primary parameters depend on the character of the pollution and the analytic possibilities of the automated system, but they also include the most simple determinations. Taking into account that relationships among the parameters change over time, mathematical models for their determination should be adaptive and all characteristic parameters for a given water body and their relationships should be measured.

An important but often insufficiently used way of obtaining warning of an EHP is by out-of-system notifications such as messages from industrial enterprises (about the accidents), sanitation and fishery services, public organizations, and citizens.

In addition to an immediate discovery of an EHP because of a change in the chemical composition of the water and in the state of the ecosystem, an indirect method of indentifying an accidental discharge may also be employed. This method uses a tracer, i.e. an ecologically harmless substance which possesses characteristics allowing it to be a detected easily at low concentrations which enters the water body together with the accidental discharge. Fluorescence is one substance with such characteristics. The use of tracers presents a number of problems, however, the basic ones being:

(a) The design and manufacture of a way to inject the tracer into the water body only when accidental discharges occur and at no other time;
(b) The creation of a reliable, simple and rapid means of tracer detection in natural waters.

A preliminary analysis of different means of tracer detection shows that the most promising is a laser on board an observing aeroplane or ship.

It must also be noted that the use of tracers is also far from universal. They may be employed to monitor discharges of large pollution sources, but basic pollution sources are often too numerous and too small to warrant their use. Moreover, they cannot be used to monitor non-point discharges.

Thus, reliable information about an EHP may also be secured through use of all the described methods. In developing a system, special attention should be paid to the time it takes to operate from the moment an EHP appears.

Only remote-sensing methods are able to determine the spatial parameters of the discharge in the water body. Where these are not available, it is necessary to carry out additional work to outline the EHP-affected zone after the appearance of the pollution. As a rule, this work may be combined with the assessment of the impact and intensity of the incident upon the ecosystem (using methods of rapid biotesting, for example) and the preliminary identification and EHP-level assessment. In the case where an EHP is observed by remote sensing, this should be combined with ground-based assessments for the detection of the spatial parameters.

From knowledge of the spatial parameters and taking into account the hydrological and meteorological data, a preliminary prediction of the spread of the pollution plume is made on the basis of which a warning about the EHP must be issued.

In comparison with remote-sensing methods, ground-based methods and

139

similar means are rather less effective for rapid determination of the spatial parameters of an EHP. Vital to this question is the need to perfect modelling of pollution spread in rivers and streams. This is a difficult problem and additional information is required.

In spite of the help that remote sensing provides in the preliminary identification of an EHP, this method is not the definitive answer due to its low resolution, the repeat cycle of the satellite and other technical difficulties which require high levels of pollution before satisfactory observations can be made. Methods have been developed which use indicator papers, visual colorimetry, ionometry and so on, along with stationary apparatus and analyzers installed in mobile hydrochemical laboratories (MHCL). At the same time that the preliminary identification an level assessment of the pollution are being made, the MHCL personnel take samples and transport them to a stationary hydrochemical laboratory (SHCL) for complete analysis.

The composition of the EHP, and the prediction of its spread and its consequences may be done on the basis of a preliminary prediction of the spread of the pollution plume and assessment of the pollution character; recommendations for the development of short-term measures to eliminate or decrease the pollution can also be made at that time. The aim of determining the characteristics of the pollution and predicting its spread and consequences is to restore the water body and the affected ecosystem to its previous state or to a new stable level.

Finally, the information received by the monitoring system should be used to improve its operation and the separate functions that it performs.

The monitoring system described for the real-time monitoring of dangerous ecologo-toxicological incidents in water bodies should determine the time scale for the operation of the system and allow timely management for the elimination of the accidental pollution and its consequences. Such a system may also be created on the international level for shared water bodies located on the territory of several countries.

The creation of such systems on the large European rivers (Danube, Rhine and others) is a pressing current issue. However, their establishment is possible only with the co-operation and efforts of all the countries concerned. Organizational and political problems may cause difficulties (for example, countries situated downstream would be more interested in an effective control system). These questions have therefore been referred to the relevant regional co-operation and development organizations where agreement on such important matters may be sought.

Water Pollution Hazard – The Need for Water Quality Monitoring as the Task for Hydrometeorological Services in Poland

═══

J. ZIELIŃSKI

Institute of Meteorology and Water Management (IMWM),
Warsaw, Poland

Water Resources and Water Demand

IN AREA, Poland is 312,520 km^2 and the total water surface area covers 5,000 km^2 or 1.6% of the total area of the country. Poland has about 9,300 lakes covering an area of 3,200 km^2. For example, the Masurian Lake District has 1,063 lakes, the largest having areas of about 11,000 ha. Lakes and artificial reservoirs have a capacity of 33 km^3, a large number of ponds hold an additional 1 km^3. The two most important rivers in the country are: the Odra river which has the area of the basin (Goczałkowice station) of 110,000 km^2, and the Wisła river with the area of a basin of 194,000 km^2 (Tczew station).

The average annual amount of rainfall is 597 mm, equivalent to 186.6 km^3 of water per year over the whole country. Since tributaries from outside Poland yield an additional 5.2 km^3 of water per annum, the total amount of water is therefore 191.8 km^3. The underground water resources have been estimated as 33 km^3 of water per year, for an area of 272,520 km^2, since the remaining 13% of the total area is waterless. The annual dynamic underground water resources have been evaluated as 9.2 km^3. However, rivers and streams discharge about 58.6 km^3 of water into the Baltic Sea in a mean low-flow year, and about 34 km^3 in a mean dry-weather year. Obviously, only a portion of this volume is available. About 10 km^3 is necessary as a minimum flow to maintain biological life and for sanitary reasons. Therefore, the available flow is only 24 km^3 of water. Hence, Poland belongs to the group of European countries most deficient in water resources. The country ranks as 22nd in Europe. The average annual water resources in Poland estimated on the basis of atmospheric falls and a number of population amount to 1,800 m^3 per habitant, comparing with 2,800 m^3 per habitant in Europe. The annual water demands for Poland in 1990 are anticipated as about 28 km^3 (13 km^3 in 1976), with about 5 km^3 for municipal supply (2 km^3 in 1976), and 9 km^3 for agriculture, comparing with 4 km^3 in 1976. Most water for agriculture is taken during summer months. Water demands and waste water discharges are shown in Figure 1. The available water volume compares unfavourably with the water demand anticipated for 1990.

Legislation and Administrative Aspects

Poland has a 60-year-old history of legislation for water pollution control. The Water Quality Act was issued in 1922 and revised in 1962. At present, the basis for legal action in the field of water protection against pollution in the country is a new version of the Water Law issued by the Polish Parliament in 1974. On the basis of the Water Law, the Council of Ministers announced in 1975 the regulations concerning the classification of waters and the determination of effluent standards, as well as financial penalties for effluent discharges which do not meet the requirements specified in the regulations.

These regulations are set up for BOD, COD, either extracts, PCB various metals, pesticides, etc. The following classes of surface water were established:

– Class I waters are those used in municipal and food processing purposes and for salmon fish growth;
– Class II waters are intended for use as recreational waters, including swimming, and for growth of fish other than salmoidae;
– Class III, the lowest class of waters that allows only for their use as industrial supplies and for irrigation purposes.

Water quality standards are tailored to meet the appropriate use of surface waters. Permissible concentrations of selected constituents for different classes are shown in Table 1, as an example. In addition, the following provisions are laid down by the Water Law: industrial plants and other operations which discharge wastewaters into water or onto land are obliged to construct, maintain and utilize waste water treatment facilities. Without simultaneous operation of waste water treatment systems no industrial plant or any other plant from which waste water is discharged should be started up. To maintain a waste water discharge it is required to obtain a permit. In order to promote administrative measures for overall conservation of the environment, the Ministry of the Environment Protection and Natural Resources was established with environmental offices on a voivodship (prefecture) level.

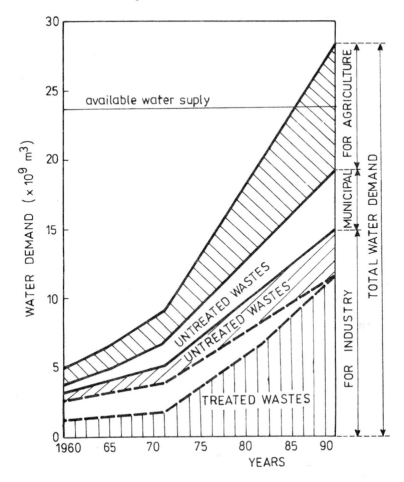

FIGURE 1 *Water demand and wastewater discharge in Poland*

Water Quality Surveillance System

The water quality surveillance system should be composed of conventional and automatic monitoring stations. The conventional monitoring with manual sampling at the voivodship (prefecture) level was established in 1957. At present, the country is covered by a network of about 3,000 conventional stations and established cross-sections located along 30,000 km of streams. The sampling frequency depends on the purpose for which data are recorded, and it ranges from a minimum of once every two months up to daily sampling at some points. Water sampling is performed simultaneously with rate-of-flow measurements. For continuous monitoring of the Odra and Wisła rivers, the automatic water quality monitoring stations (AWQMS) have been installed at

143

TABLE 1 *Examples of permissible concentrations of some pollutants in surface freshwaters according to the Polish Water Law*

Parameter	Unit	Water class		
		I	II	III
DO	mg O_2/dm^3	6	6	4
BOD$_2$	mg O_2/dm^3	4	8	12
COD	mg O_2/dm^3	40	60	100
Saprobic index		oligo to betamezo	betamezo to alfamezo	– alfamezo
Chlorides	mg Cl/dm^3	250	300	400
Sulphates	mg SO_4/dm^3	150	250	250
Hardness	$mval/dm^3$	7	11	14
Dissolved solids	mg/dm^3	500	1000	1200
Suspended solids	mg/dm^3	20	30	50
Temperature	°C	22	26	26
N—NH$_4$	mg NNH_4/dm^3	1.0	3.0	6.0
N—NO$_3$	mg NNO_3/dm^3	1.5	7.0	15
N—organic	mg $Norg/dm^3$	1.0	2.0	10
Total iron	mg Fe/dm^3	1.0	1.5	2.0
Manganese	mg Mn/dm^3	0.1	0.3	0.8
Phosphates	mg PO_4/dm^3	0.2	0.5	1.0
Cyanides	mg CN/dm^3	0.1	0.02	0.05
Phenols	mg/dm^3	0.005	0.02	0.05
Lead	mg Pb/dm^3	0.1	0.1	0.1
Mercury	mg Hg/dm^3	0.001	0.005	0.01
Copper	mg Cu/dm^3	0.01	0.1	0.2
Zinc	mg Zn/dm^3	0.01	0.1	0.2
Cadmium	mg Cd/dm^3	0.005	0.03	0.1
Chromium	mg Cr/dm^3	0.05	0.1	0.1
Total heavy metals	mg/dm^3	1.0	1.0	1.0

important cross-sections, connected mostly with water intakes. The AWQMS are either in laboratory buildings or on barges. The automatic network, one of the first in Europe, has been operating since 1968. It was organized under the auspices of the World Health Organization, and initially was based on imported monitoring equipment. At present, domestic automatic monitors (Aquamers) are being used. These monitors are capable of measurement and teletransmission of such parameters as: water temperature, pH, conductivity, dissolved oxygen (DO), chlorides, turbidity, water level and meteorological data. Additional parameters will be added after suitable sensors are developed. However, it should be stressed that the scarcity of automatically measurable parameters limit the efficient application of AWQMS. The development of automatic measuring devices for other important parameters such as heavy metals, organo-chlorine compounds, oxidized nitrogen, soluble organic content and others is greatly needed. At present, automatic stations are also used for manual bioassay tests and fish tests. The future surveillance development programme calls for the Basic Water Quality Monitoring Network (BWQMN), based on a limited number of stations (with extensive measurements as supplementary to the operational water quality surveillance system (Zieliński et al., 1988).

Background Data and Objectives of Monitoring System

The existing conventional monitoring system does not give a comprehensive picture of water quality due to the varied measurement periods, low sampling frequencies and insufficient number of water quality parameters. The proposed basis water quality monitoring network (BWQMN), based on a limited number of selected stations (initially 10 stations increasing to 48) will be supplementary to the local conventional system.

Specific objectives of the basic water quality monitoring network are:

(a) determination of water quality baselines and trends all over the country and not at the specific sewage discharge points;
(b) establishing the relationships that would enable the prediction of water quality according to hydrological conditions;
(c) assessment of pollution loads from various areas of different types and management intensity;
(d) assessment of total load discharged into the Baltic Sea by the main Polish rivers, the Wisła and Odra;
(e) establishing a basis for comparative pollution analyses of watercourses in Europe and other continents;
(f) promotion of the assessment of the environmental impacts of man's activities on water resources.

Hence, the state monitoring of surface water quality will be conducted in two systems, (1) *Basic system,* and (2) *Repering system.*

The basis system is aimed at the monitoring of pollution in main rivers at cross-sections that close the catchment areas. Forty-eight basic cross-sections are planned, including 10 balanced cross-sections equipped with automatic stations.

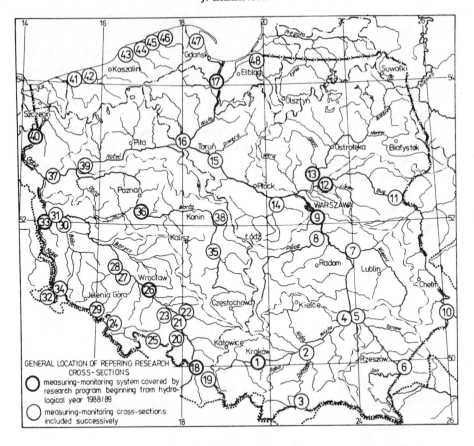

FIGURE 2 *General location of repering research cross-sections*

The cross-sections are as follows (Figure 2):

No. 1 on the Wisła river upstream of Cracow;
No. 9 on the Wisła river near Warsaw;
No. 17 on the Wisła river near Kiedzmar;
No. 12 on the Bug river upstream of Zegrzyński Reservoir;
No. 13 on the Narew river;
No. 18 on the Odra river at Chałupki;
No. 26 on the Odra river near Wrocław;
No. 40 on the Odra river near Szeczecin;
No. 33 on the Nysa Łużycka river;
No. 36 on the Warta river.

Five indices will be measured continuously (temperature, pH, dissolved oxygen, conductivity, chlorides) and for the rest about 100 measurements are carried out annually (approximately two each week). Samples for other indices

146

will be collected automatically and frozen, and then marked manually by means of a special device. If automatic stations have been introduced, then each station in the base network (10 cross-sections) will be equipped with the water level recorders. The operation of the base network will be supervised and analyses carried out by the national Meteorological Service (IMWM).

The repering system aims to monitor water quality in those rivers which are of the greatest importance from the economic point of view. One thousand and twenty-six measurement points on 55 rivers monitored each year are planned. The planned sampling frequency is twice a month. From rivers polluted by seasonal industries (e.g. sugar industry) samples will be collected four times a month during the campaign season. Basic indices marked at each measurement point are: temperature, dissolved oxygen, BZT_s, ChZT-14, chlorides and others.

The operation of the repering network will be supervised by local environmental authority laboratories, which exist in each voivodship (prefecture). The analyses will also be made by those laboratories. All data will be collected and processed in the Central Data Bank of the national Meteorological Service.

Measurement Programme for the Monitoring Stations

At all stations various water quality parameters (together with the rate of flow) will be measured in order to assess the water quality on the annual bases. Sampling frequency at each main station will be once or twice a week (not less than 100 measurements per year). The water quality parameters to be measured at each station are:

- Water temperature;
- pH;
- Dissolved oxygen;
- Conductivity;
- Chlorides;
- Total hardness;
- Sulphates;
- Colour;
- Odour;
- Turbidity;
- Biochemical oxygen demand (BOD, five days);
- Permanganate value (PV);
- Chemical oxygen demand;
- PCBs;
- Phenols;
- Cyanide;
- Detergents;
- Sodium, manganese calcium, potassium;

- Fe, Zn, Cr, Cu, Pb, Cd, Hg, Ni;
- Phosphorus (total and dissolved);
- Phosphates (total and dissolved);
- Ammonia-N, nitrate-N, nitrite-N, total-N;
- Total bacteria;
- Coliforms;
- Saprobic index.

In the case of the 10 automatic monitoring stations to be installed, the first five parameters will be measured continuously.

Main Activities

The project will comprise the following tasks:

First Phase

- Selection of appropriate exact sites for representative location of cross-sections;
- Final selection of water quality parameters and sampling frequency;
- Selection and unification of analytical methods applied in all laboratories;
- Training of technicians, transport and sample storage organization;
- Establishing a detailed research programme for the project and developed of reporting forms.

Second Phase

Intensive field and laboratory studied of the water quality carried out at the monitoring stations.

Simultaneously with the first and second phases, the following specific research studies will be carried out:

- New analytical methods for various substances and their implementation in routine work of laboratories;
- New quantitative methods for water quality assessment;

Computerized Data Analyses

The river monitoring network provides a large number of observational data. The data will be analyzed by means of statistical method based on the assumption that at a given cross-section some correlation exists between the pollutant concentration and the rate of flow (Figure 3) (Environmental Pollution Abatement Centre, 1976). Basic types of curves representing this correlation are applied. The shape of the curve depends on various factors such as the level of water pollution, the type of pollutant, hydrological

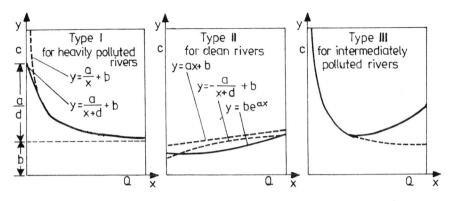

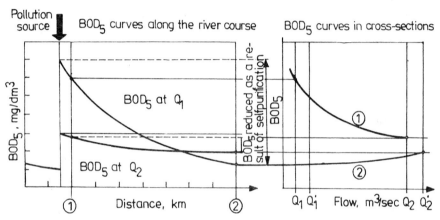

FIGURE 3 *Relationship between concentration of pollutants and rate of flow*

characteristics of the river, its self-purification capacity, the distance between monitoring stations, etc.

A final intrepetation will be based on such hydrochemical characteristics, and the river overall classification will be performed after all measured water quality constituents have been compared with the standards. A compendium of hydrochemical profiles for major rivers and streams will be prepared each year.

Use of Outputs

The main outputs of the project will be presented in the form of official reports. The data obtained from the basic monitoring network would be utilized for:

(a) The determination of water quality baselines and trends necesary for strategic planning decisions;

(b) The assessment of total pollution loads discharged into the sea by main rivers;

(c) The introduction of the effective, preventive measures.

In order to attain satisfactory water quality goals, a meaningful and economical water quality data monitoring system is mandatory. An additional benefit is the establishment of a basis for comparative pollution analyses of water courses in Europe and other continents.

Cost of the Project

Setting up of the basic monitoring system (BWQMW) will require the setting up of four additional main laboratories (49 laboratories already exist) that have to be equipped with modern devices such as: spectrophotometers (atomic absorption); spectrophotometres VV-VIS, ICOP, IR; Technicon autoanalyzers; Backman carbon analyzers; gas chromatographs; oxygen analyzers.

The cost of the project for 48 stations will be about 6,000,000 Polish Zloties (equivalent to US$ 12,000,000).

REFERENCES

ADAMCZYK, Z. *et al*. (1978) Simple mathematical model of quantitative and qualitative process occurring in the stream channel for water distribution control. *Proc. Baden Symp. on Modelling the Water Quality of the Hydrological Cycle*. IAHS-AISH Publ. No. 125

ENVIRONMENTAL POLLUTION ABATEMENT CENTRE (1976) Environmental Protection UNDP/WHO Project POL/RCE-001. *Final Report*, Ślask Publishing Company

GROMIEC, M.J., LOUCK, P.D. and ORLOB, G.T. (1983) Stream quality modelling. In: *Mathematical modelling of water quality: streams, lakes and reservoirs* Orlob, G.T. (Ed.) John Wiley & Sons, Chichester

GROMIEC, M.J. (1983) Biochemical oxygen demand – dissolved oxygen: river models. In: *Application of ecological modelling in environmental management, Part A* Jorgensen, S.E. (Ed.) Elsevier Scientific Publishing Company, Amsterdam

GROMIEC, M.J. and ZIELIŃSKI, J. (1988) *River water quality assessment and modelling in Poland* Institute for Meteorology and Water Management, Warsaw, Poland

MANCZAK, H. (1973) Statistical method for estimation of water quality monitoring data and hydrochemical profiles of rivers. United Nations Econ. Soc. Council, CES/SEM

ZIELIŃSKI, J. *et al*. (1988) *State monitoring of surface flowing waters*. Institute for Meteorology and Water Management, Warsaw (in Polish).

Consequences of the Sandoz Accident for the Biozonosis of the Rhine

Report of Results from the German Commission for Prevention of Pollution of the Rhine

—

KARL HANS HEIL

*Hessiches Ministerium für Umwelt und Reaktorsicherheit,
Wiesbaden, Federal Republic of Germany*

Introduction

TWO YEARS AGO, on 1 November 1986, Central Europe was shattered by an environmental catastrophe in the Basle area whose physical, chemical and biological consequences for the area and especially for the River Rhine were immense. Two years later the direct consequences seem to have mostly been overcome. The public's worries that such an accident may happen at any time again have not yet passed.

The fire catastrophe of 1 November 1986 in the Schweizerhalle district of Basle emphasized in one night to what extent people and the entire environment are endangered in the immediate neighbourhood of cities. With great effort the emergency and fire services were able to overcome the dangerous chemical fire. People in the immediate and distant neighbourhoods were protected from the worst. Only a few hours and days later news of the consequences became public:

- immense loss of fish stock in the upper Rhine;
- destruction of the aquatic life, especially of the Macrozoobenthon, as far as the Mainz area;
- drinking water threatened;
- endangering all uses of the water.

The accident dominated the entire public discussion to an extent unknown before. The public, already alerted by the Chernobyl accident, were acutely aware of this new environmental catastrophe.

Was this incident a warning sign to other, large and small, emitters? Did industry immediately (within days) draw conclusions and take preventive action – or was opinion one of 'this could never happen to us'? Did one or two even take advantage of the situation to emit harmful waste? In the weeks following the accident, reports of further accidents were certainly more numerous than usual.

Were these happenings coincidental, simulated actions of certain groups or were newly-aware, shocked emitters finally ready to reveal everything that usually disappeared under the cover of silence?

The Sandoz accident caused national and international political activity:

– immediately after the accident the ministers from states neighbouring the Rhine met for the first time (12 November 1986) in Zurich;
– a second conference at ministerial level, which had been extensively prepared and which presented many constructive suggestions, took place on 19 December 1986 in Rotterdam. This was the basis for the 'Action Programme Rhine' which was finally ratified at a third conference at ministerial level on 1 October 1987 in Strasbourg.

This political activity was an expression of utmost worry and constituted a serious attempt at examining the facts of the case thoroughly, and at making changes to avoid future similar catastrophes as far as possible. Political actions were marked by activity rather than mere words. I will refer to the 'Action Programme Rhine' as a definite product of these activities later. In the administrative area, widely co-ordinated, purposeful assessment of the consequences of the environmental catastrophe took place. This is true not only for the boards of the International Commission for the Protection of the Rhine (IKSR) but also for the national boards of Switzerland, France, the Netherlands and the Federal Republic of Germany (FRG).

The greatest organizational, personnel and financial expenses came from the German side. The appropriate committee of the German Commission for Prevention of the Rhine pollution (DK) the participating departments (Federal and State) which belong to this Commission, as well as the subordinate offices, the investigation authorities, technical offices (Federal and State) as well as the affected Water Authorities were constantly active.

A Warning and Alarm System was immediately set up, as well as the nationally and internationally co-ordinated 'Catastrophe Measuring and Research Programme' whose evaluation, classification and documentation demanded incomparable action from everybody. As early as the Ministerial Conference on 19 December 1986 in Rotterdam, all member states affected by the incident, as well as the IKSR, were able to deliver the first written reports about the extent of damage.

The 'German Report on the Sandoz Accident' which was established by a working committee in excellent co-operation with all appropriate Federal and

State authorities was ready by the beginning of December and contained:

- complete results of measurements within the chemical-physical survey of presence of materials relevant to Sandoz in the draining pollutants;
- a first evaluation of the Macrozoobenthon stock in the Rhine;
- a thorough ecological-toxicological evaluation;
- a medium-term measurement programme for the timeframe of one year after the accident;
- a programme for re-stocking of the Rhine, especially for eels;
- a draft for necessary research and development plans.

Other German reports covered the important issues of:

- handling of claims and liability;
- additional safety precautions for industrial installations in case of catastrophes, especially fire;
- improvement of the Warning and Alarm Systems.

All the above-mentioned German activities have definitely influenced the international discussion. They finally resulted in the 'Action Programme Rhine'. Today, exactly two years after the accident took place, the question is asked: Which activities have been carried out, and how many still remain to be done? Can we hope and expect that definite conclusions drawn from experience of the Sandoz catastrophe and the final conclusions and proposed programmes drafted by politicians, administrators and industry will have definite consequences for the life in and on the Rhine?

This is not the concern of my topic, so I am unable here to give you a status report on these activities.

I want to restrict myself to three areas:

(1) the consequences of the Sandoz accident for the quality of the water and for the biozonosis of the Rhine;
(2) present and future measuring programmes for the Rhine;
(3) the most important steps as they appear in the 'Action Programme Rhine'.

The Consequences of the Sandoz Accident for the Ecology of the Water and Biozonosis of the Rhine

As a result of fire damage to a Sandoz warehouse in the Schweizerhalle district of Basle on 1 November 1986 approximately 10–30 tons (according to German calculation) of insecticides (phosphorus acid components [Disulfoton Thiometon, Etrimphos and Propetaphos] and organic mercury compounds) reached the Rhine via the water used for firefighting, where they had a profound effect on the quality of the water as well on as fish and organic life.

Based on the 'German Report to the Sandoz Accident (with laboratory analyses) December 1986' the ministers of the states neighbouring the Rhine gave the following order to the IKSR on the occasion of their conference on 19 December 1986 in Rotterdam: to co-ordinate the already started short, medium and long-term national survey programme in reference to the consequences of the Sandoz accident, and undertake a re-stocking programme in co-operation with fishing experts.

Thereupon a medium-term research and re-occupation programme was conducted. The results of the German part of the Rhine are presented below.

Survey results during and immediately after the discharge of the pollutants (summary)

The survey results are summarized in the '2nd German Report on the Sandoz Accident: condition of the Rhine one year after (July 1988)'. Copies may be obtained through the Chairman of the German Commission for Prevention of the Rhine Pollution (DK), Post Box 30 06 52, 4000 Düsseldorf, Federal Republic of Germany.

The pollutants passed through the Rhine in the Federal Republic of Germany from 1–13 November 1986. The process of concentration of disulfoton, thiometon and the sum of the phosphorus acid components are presented in Figures 1, 2 and 3 and the maximum data of the measured pollutant concentration in Table 1.

Influence on the fish

From the point of entering (Rhine km 159) to beyond the Lorelei (Rhine km 560) according to the survey results obtained immediately after the accident, the entire stock of eels in the Rhine, including its branches, was destroyed. Extensive damage was noted to the Rheinland Pfalz/Nordrhein Westfalen State Line (approximately Rhine km 649). In Nordrhein Westfalen no loss of eel stock was noted. In addition to that, stocks of other fish species (such as the Äsche salmon species, pike, pike-perch and others) were damaged in Baden-Württemberg.

Consequences for the fish nutrients (Macrozoobenthon)

In the vicinity of the accident a total loss, with only few exceptions, of fish nutrients was noted. Also in the remainder of the Rhine between Basle and Neuenburg severe damage was noted. However, in the lower part of the remaining Rhine between Neuenburg and Breisach no damage was noted as a result of the considerable subsoil inflow, with the exception of a partly reduced stock of Köcherfliegen (water fly larvea = Trichoptera) and flea crayfish.

Also in the side arm between Basle and Breisach reductions of stocks, up to a total loss of some species, was noted. In the regions between Breisach and

TABLE 1

Highest values of measured pollutant concentrations

Location	Rhine (km)	Date (November 1986)	Time	Disulfoton (µg/l)	Thiometon (µg/l)	Etrimphos (µg/l)	Propetamphos (µg/l)	Oxadixyl (µg/l)	Parathion-ethyl (µg/l)	Mercury (µg/l)
Märkt	173.00	1	15.15	600	500	50	100	80	—	12
Wyhl	244.35	2	16.45	107	23	10	6	37	1	2.6
Gambsheim	310.00	4	00.00–03.00	73	15	4	2	32	0.5	0.6
Maximiliansau	362.00	4	12.00–24.00	24.6	10.6	3.1	1.1	11.5	0.4	—
Ludwigshafen	428.00	5	10.33	30.3	14.4	3.1	1.0	12.3	0.4	0.4
Mainz	498.00	6	09.00	18.3	8.3	2.6	3.4	—	0.4	—
Koblenz	590.00	7	04.00	—	—	—	—	—	—	0.2
Neuwied	609.00	7	09.00–10.00	11.8	3.9	1.3	0.6	6.9	0.3	—
Bad Honnef	640.00	7	14.00–18.00	8.9	3.5	1.1	1.0	—	0.1	—
Düsseldorf	734.00	8	08.15	5.7	2.2	0.7	0.6	—	< 0.1	—
Lobith	862.00	9	09.00	5.3	2.0	—	—	—	—	0.22

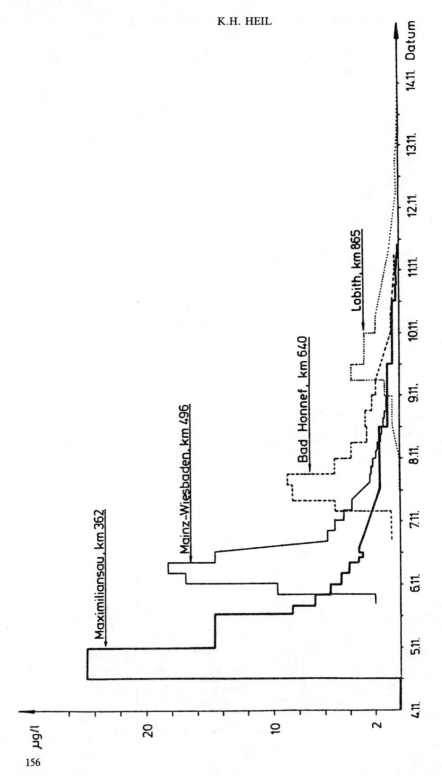

FIGURE 1 *Concentration of disulfoton against time (x axis shows date)*

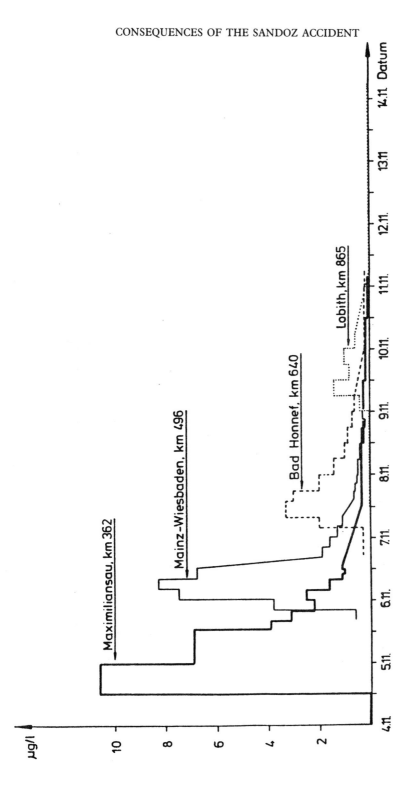

FIGURE 2 *Concentration of thiometon against time* (x *axis shows date*)

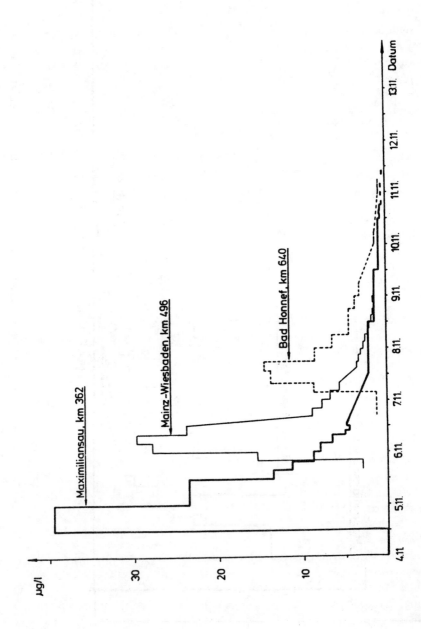

FIGURE 3 *Concentration of all phosphoric components against time (x axis shows date)*

Bad Honnef regionally, both to the left and right of the Rhine, differing but definite damage was noted. Only from Bad Honnef to the Dutch border did the Rhine show no more signs of damage.

Effect of the pollutants

The pollutants which leaked into the Rhine due to the fire show a moderate to strong virulence in mammals; they are poisonous for fish in the quantities of a few mg/l and affect the inferior water organisms fatally at a few mg/l. Phosphorus acid components paralyze the nerve centres by blocking the enzyme cholinesterase. A list of the Eco-toxicological data can be found in the German Sandoz Report.

Medium–term survey programme after the accident

According to the recommendations of the German Commission, a Macrozoo-benthon count of the banks of selected parts of the Rhine and its side arms was taken monthly during 1987 by the appropriate State authorities. The Federal Office of Hydrology in Koblenz sampled the Rhine at 15 points to test the 'Macrozoobenthon on the bottom of the Rhine'

At 23 places along the Rhine, and its branches, fish (eels, non-predatory and predatory fish) should be taken and tested for substances relevant to Sandoz; mercury, atrazine, HCH, HCB and PCB.

Survey of the Rhine along its banks

Because of high and changing water levels the analysis in the most severely affected areas, Baden-Württemberg, Hessen and Rheinland Pfalz, could be conducted only sporadically. From mid-August to September 1987 it was possible to conduct at the same time an inventory of the re-settling for the entire Rhine. The results have not been evaluated.

The German Sandoz Report shows a comparison of the evaluation results of the strength of Macrozoobenthon directly after the accident and the results of August-September 1987. The results show that in many species *no more damage can be detected*. Practically all species present before the Sandoz accident can now be seen again, even though the number of single individuals is in many cases reduced.

In detail the following can be seen:
The stretch of the Rhine from Schweizerhalle to above the mouth of the Neckar, which after the accident was marked by strongly reduced settling through to total loss of single species, showed by the end of 1987 less reduced settling and, in parts, no more signs of damage. An exception are the Köcherfliegen (water fly = Trichoptera), which were especially affected by the accident. Definite deficiency can still be seen here.

From below Germersheim to Lorch different results can be seen on the left and right of the Rhine, which mostly belong to the categories 'no more detectable damage' or 'small reduction'. Compared with the results from the end of 1986 (reduced settling to total loss of single species) a significant positive change can be seen.

From below Lorch to Bimmen, close to the German-Dutch border, no difference can be observed between the conditions immediately after the accident, and those of Autumn 1987. Practically no effect of the pollution drainage was noted, so no detectable damage was noted either after the accident or in later times. The surveyed side arms show similar results. The stock of fish nutrients which was reduced by the Sandoz accident is mostly restored.

Survey of the bottom of the Rhine

Extending the initially planned measuring programme, which entailed two tests in 1987, the Federal Office of Hydrology conducted a further profile test of the entire Rhine in different seasons at 15 testing areas in connection with the BMU Research and Development Programme 'Survey of fauna at the bottom of the Rhine for determination and evaluation of the damage to the biozonosis'. The last inventory was done in October 1987, corresponding with the survey of the banks.

It proved to be advantageous for the survey of the bottom of the Rhine to be conducted independently of the Rhine water level and therefore give detailed information on the variety of species and density of individuals of the biozonosis during the entire year of 1987. The interpretation of the results with regard to the Sandoz accident is aggravated by the fact that there are no results available of the methodically similar reference tests carried out on the Upper Rhine, the critical area, before the accident. Therefore it is not possible to make strictly valid comparisons to the status before the accident, as was done for the Lower Rhine below at Bonn. By comparing the results obtained directly after the accident and those of later tests, accurate statements regarding development tendencies of single populations can be derived.

The first bottom survey in 1986, done in the vicinity of the accident, showed when compared with the results of the right bank survey, that the population of the different insect and crayfish species, especially Hydropsyche and Gammarus species, had strongly diminished, while the snail, mussel, leech and strudelworm (Turbellaria) stock had evidently not diminished. Downstream the occupation improved whereas the strongly damaged crayfish and insect species of the accident neighbourhood area already showed in the dam area of Iffezheim a lower to medium, and below the mouth of the Main a higher individual density.

Comparing the results obtained in December 1986 with those of 1987 a low-to-strong increase in water fly (Trichoptera) and flea-crab stock was observed, while other organism groups showed no increase but a partial decrease, probably due to the season. Below Koblenz no changes could be noted compared with the results from the end of 1986. Therefore it may be concluded that no damage to the Benthal fauna was caused by the Sandoz poisons.

Details of this survey may be obtained from the Federal Office of Hydrology, which has established extensive documentation of the survey.

Findings on the development of fish stocks

The fear that the fish fauna could have been damaged for years and that the positive development tendencies of the past years might have been interrupted, have not proven to be true according to the survey programmes done. Comparison of the species before and after the accident show that at least none of the species existing earlier have disappeared completely. The formerly dominating species such as roach, bleak, bream and river perch (Blicca) are still present.

Regeneration of eel stock also took place sooner than assumed. Further re-occupation is necessary, in order to achieve a *complete* re-occupation within a time-frame of 7 to 8 years.

Re-occupation programme for fish stocks

In 1987 the states of Baden-Württemberg, Rheinland-Pfalz and Hessen put into the German part of the Rhine between Basle and Mannheim some 400 kg of eel brood (= 1.2 million elvers) and into the Hessen part of the Rhine they have so far put 360 kg of eel brood. The re-stocking with eels is planned to continue. It is assumed that the re-occupation of eels will take up to 8 years – surveys of the coming years will show whether this proves to be correct.

Present and Future Measuring Programmes for the Rhine

Irrespective of the above-mentioned medium-term measuring programme resulting from the Sandoz accident, the states along the Rhine conduct intensive measuring programmes which I would like to cover briefly:

– continuous biological surveys according to the Saprobien system. The results for the individual States as well as for the entire Federal Republic of Germany are published on a regular basis;
– German Rhine Measuring Programme: at seven fixed measuring stations of the Rhine as well as at eight measuring stations on its most important branches, approximately 40 parameters are continuously measured and analyzed, documented and published in annual tables;
– three stations of the German measuring programme are part of the international measuring programme, where results are also published in annual tables;
– each of the states along the Rhine completes these sub-regional programmes through special surveys of organic micro-pollution.

In the field of organic micro-pollution all measuring, analysis and evaluation procedures have to be updated. Possibilities have to be found in the area of mutual national and international activities in the framework of pilot surveys, orientation measurement programmes and research and development plans to improve the count and evaluation of these materials. Aside from measuring in the watery phase, testing of suspended matter and sediment surveys of fish, fish nutrients and other aquatic organisms will be important. Measurements are supposed to show on a monthly basis the current condition of the waters, developments resulting from actions taken, and also function as a warning system in case of sudden pollution. Therefore the German Commission for Prevention of Rhine Pollution presented to the International Commission for Protection of the Rhine two research and development plans to combat pollution in order to obtain further knowledge for water control:

(1) Research and development plans for testing a measurement procedure for certain organic micro-pollution under consideration of single and sum parameters.

   This research plan will be stated on an international basis with participation of Switzerland, France, the Netherlands and the Federal Republic of Germany on 1 November 1988 by the Engler-Bunte Institute at Karlsruhe.

(2) Research and development plans 'Research and testing of biotest-procedure for determination of acute toxic effect of peak load in the Rhine'.

   This plan, which is being prepared by the Federal Republic of Germany for the entire German part of the Rhine, is supposed to begin soon.

Both research plans should run for three years and could possibly deliver results within this time. In consideration of the improvement of *water control*, surveys for an optimation of the drainage control should be conducted. Here, too, the Federal Republic of Germany is preparing suggestions for further proceedings and a possible definition. Especially important for this research programme will be the participation of the main emittants.

(3) A report of the consequences of the Sandoz accident would be incomplete without mentioning, at least in general terms, the Action Programme Rhine, which was politically initiated as a result of this accident.

According to the resolutions of the Ministerial Conference of 19 December 1986 in Rotterdam, the International Commission for Protection of the Rhine established the 'Action Programme Rhine', which was ratified at the Strasbourg Ministerial Conference on 1 October 1987. With the help of this programme the following should be obtained by the year 2000:

– re-settling of previously existing high-level species (e.g. salmon) in the Rhine;

– that the use of the Rhine as drinking water will be possible in the future;
– that the sediments be cleared of poisonous materials.

The main goals of the appointed activities are:

– an accelerated reduction of strain by direct and diffuse drainage;
– a reduction of accidental dangers;
– an improvement of the hydrological, biological and morphological conditions.

The realisation of the Action Programme Rhine should be completed in three phases. In the first phase, up to 1989:

– priority materials and the relevant industrial firms and fields to be determined;
– national Emission Registers for these materials to be established;
– an estimation to be done of the possible reduction of the drained pollutants;
– technical concepts for the improvement of hydrological and morphological conditions to be developed;
– an estimation of cost to be established.

In the second phase, until 1995 especially:

– the priority materials are to be reduced drastically in accordance with the technical standard (possibly 50%);
– a minimum control programme for drainage control to be established;
– action to be taken for higher security of industrial installations;
– action for hydrological and morphological adaption to be conducted;
– a concept to be established for reduction of strain by diffuse sources.

In the third phase, up to 2000:

– additional action to be taken should the action undertaken during the first two phases not have reached their goals.

You can see from the above explanations, that the consequences of the Sandoz accident have contributed to the necessity to keep water clean becoming once again public concern.

It is hoped that this political impulse, which was strongly welcomed by the experts, will also find support with those institutions who have to provide the necessary financial and organizational basis.

Hydrological Aspects of Pollution of the Rhine

—

J. VAN MALDE

President, International Commission for the Hydrology of the Rhine Basin (CHR), Rijkswaterstaat, DGW, The Hague, the Netherlands

Some General Information on the Rhine

ACCORDING to world standards the Rhine (Figure 1) is not really a large river: with its length of 1,320 km it only ranks 72nd on the list of the world's largest rivers. In Europe eight rivers have a larger catchment area than the Rhine, ranging from the Volga (with a basin nearly 7.5 times as big as the Rhine basin) via the Danube to the Vistula (corresponding ratios: 4.35 and 1.05). Yet the Rhine is Europe's most important navigation route. This is not only due to human efforts, but also to:

(1) auspicious geographical conditions, one being that the river flows from the Alps to the North Sea, the latter offering excellent possibilities for human activities: fishing, shipping and – recently – extraction of oil and natural gas;

(2) favourable hydrological conditions: the river has a relatively high mean discharge, whilst in the period March–August the melting of snow in – mainly – the Alps on the average neutralizes to a quite considerable extent the then reduced runoff in the remainder of the Rhine basin (Figure 2). This phenomenon restricts not only the lengths of periods of low discharges but often also the extent of the decrease of discharge. Table 1 presents a few data on the regime of the Rhine (Lobith is the village along the Rhine closest to the German-Dutch border). Chapter A5 of the Monograph on the Rhine Basin (CHR/KHR, 1978) presents a review of the regime of the Rhine and its tributaries.

The evolution of the Rhine to a first-rate navigation route was accompanied by the development of many large, and even huge, industrial areas along the

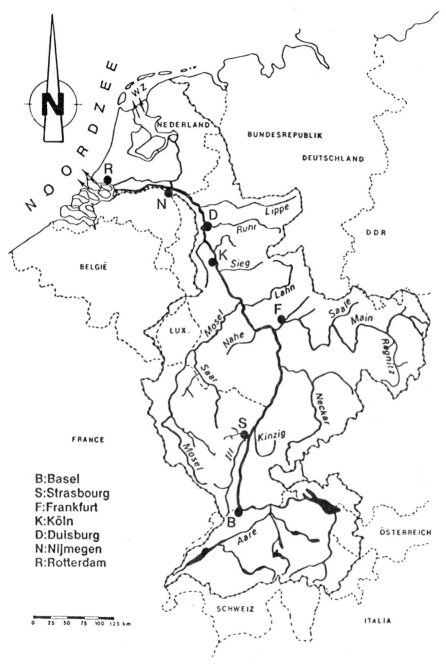

FIGURE 1 *The Rhine basin* (*Schweiz*, Switzerland; *Österreich*, Austria; *Bundesrepublik Deutschland*, German Federal Republic; *Noordzee*, North Sea; *WZ*, Waddenzee, or Wadden Sea; *Köln*, Cologne)

TABLE 1 *Some hydrological data on the Rhine*

Station	Catchment area (km^2)	Discharges (m^3/s)			Ratios	
		Highest	Mean	Lowest	Highest: Mean	Lowest: Mean
Basle	35,900	4,330	1,035	200	5.8	0.19
Lobith	160,800	13,000	2,200	620	5.9	0.28

river and its tributaries. Roughly speaking two periods of industrialization may be distinguished, being 1870 to 1945 and the period of accelerated development 1945 to the present day. There is a large variety of industries, etc.: mines, refineries, chemical plants, other industries, nuclear power stations. Small wonder in view of all these activities that the Rhine basin (185,000 km^2) is densely populated, being inhabited by some 50 million people. However, the average population density of 270 inhabitants per km^2 is highly surpassed in industrial areas; in the Ruhr area for instance, the major industrial area of the Rhine, the average density is 875 inhabitants per km^2.

The Rhine is an international river, its basin lying (apart from some small areas) in Switzerland, Liechtenstein, Austria, the Federal Republic of Germany (FRG), France, Luxemburg and the Netherlands; before 1870 the number of riparian states was even greater. In 1831 this complicating factor led to the conclusion of the first treaty between the Rhenish riparian states in order to ensure free navigation on the river (Act of Mainz, substituted in 1868 by the Act of Mannheim). With the Act of Mainz the Central Commission for Rhine Navigation was founded, which still exists. It was not until after World War II that other international Rhine commissions were formed, of which may be mentioned the International Commission for the Protection of the Rhine against Pollution (ICPR, 1950) and the International Commission for the Hydrology of the Rhine Basin (CHR, in 1970).

In 1963 the ICPR was based on the Berne Agreement, which is recognized by International Law.

Human Reshaping of the Rhine

Chiefly for navigation purposes man has carried out many riverworks (mainly in the period 1800–1977), ultimately resulting in a radical reshaping of the Rhine into a one-channel summerbed with flood plains over considerable lengths, which not uncommonly are the narrow remains of the 'original' ones. All its main tributaries have been canalized, as is the case with the Rhine itself

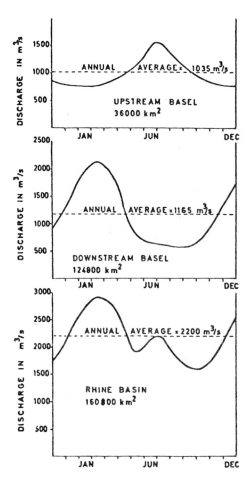

FIGURE 2 *Mean discharges in the Rhine throughout the year from Basle to Lobith*

over some 100 km (from about 60 km upstream of Strasbourg to approximately 40 km downstream of it), moreover the river is bypassed over the adjacent upstream reach (roughly speaking from Basle to halfway to Strasbourg) by the Grand Canal d'Alsace, lying within French territory. All these works have served their purpose well as may be illustrated by the fact that Duisburg in the Ruhr area is the world's largest inland port, whilst Rotterdam, situated at the mouth of the river, is the world's largest seaport.

A point of special interest is that in its delta in the Netherlands, huge civil engineering works (Zuiderzeeworks and Deltaworks), accomplished separately from the normalization of the Rhine branches, were built primarily to protect those parts of the country which are vulnerable to flooding by storm surges.

167

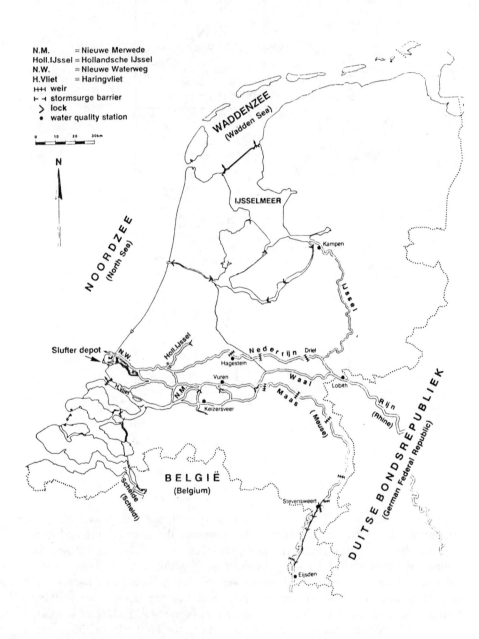

FIGURE 3 *Main water infrastructure of the Netherlands*

TABLE 2 *IJssel discharges (m^3/s) without and with operating weir Driel*

Q_L	$\hat{Q}_{ij}$	ΔQ_{ij}	Q_{ij}
2,200	310	5 (1%)	315
1,400	185	100 (55%)	315
800	80	60 (75%)	140

Nevertheless they serve other important interests, like navigation and water-management, in an excellent way as well. In the western part of this country (Figure 3) the two main Rhine branches Waal and Nederrijn share with the Meuse the two outlets left after the completion of the Delta works. Of these two outlets only one (the Nieuwe Waterweg) has a free passage to the sea; some 15 km further south a large battery of sluices in the Haringvliet is used also to discharge river water and moreover to regulate the distribution of the discharge between the two outlets, thereby protecting a large fresh-water area. The third and smallest Rhine branch (the IJssel) flows into the other main Dutch freshwater reservoir, the IJsselmeer (meer=lake), which is a part of the former Zuiderzee and discharges its surplus water in the Wadden Sea via two large sluices in the enclosure dam (30 km long) of the Zuiderzee.

When Q_L (Rhine discharge at Lobith) is lower than the mean value (2,200 m^3/s) the operation of the most upstream weir Driel in the canalized Nederrijn provides for an extra IJssel discharge ΔQ_{ij} varying with Q_L, thus increasing the 'undisturbed' IJssel discharge $\hat{Q}_{ij}$ to the real value Q_{ij}. This extra discharge ΔQ_{ij} first increases with decreasing Q_L, but decreases when Q_L falls below 1,400 m^3/s (see Table 2).

The Pollution of the Rhine and its Sources

The use of Rhine water is by no means restricted to navigation and water management (agriculture, control of salt intrusion, etc.): it is also of great importance to waterworks, industries (process and cooling water) and recreation, whilse these days the ecological values and possibilities of the river are widely recognized. All these uses of the river and its ecology are severly harmed by pollution.

In fact for a great number of centuries the Rhine has been polluted from the many old towns situated by the river: Basle, Strasbourg, Frankfurt, Cologne, Nijmegen, etc. For a long time this did not really constitute a threat as nature could cope – not until the 19th century did pollution become a burden. An ever-increasing burden, too, as the old practice of simply dumping the majority of waste was continued despite its ever expanding bulk and changing composition. By 1970 the Rhine was not only Europe's most important navigation route, but also its biggest open sewer and the condition of the river had become fully unacceptable. The sources of this pollution, consisting of a very great number of substances and chemicals, were the towns and urbanized regions, agriculture mostly run on modern lines (diffuse sources) and the industries.

It generally makes sense to distinguish between day-to-day pollution and polluting accidents; as this holds for the hydrological aspects of pollution too, the two forms of pollution will be treated separately.

Hydrological Effects of Day-to-Day Pollution

The relevant hydrological impact of day-to-day pollution is the adherence of certain pollutants to silt particles; of these contaminants heavy metals and organic micropollutants (PCBs, HCB, PAHs) are of special relevance. Under favourable conditions (little or no current) these silt particles settle; this occurs on flood plains, in harbours, in dead zones, at the turning of the tide in the lower reach of the river, in lakes, etc. This settling of contaminated silt has been going on for many years and has turned out to be an insidious calamity as will be shown.

The development of the pollution is illustrated (Figure 4) by the data on concentrations of heavy metals in silt deposits in the Nieuwe Merwede (the southern tidal branch of the Waal in the Netherlands) over the period 1900 to 1985 (this graph and Figures 5–7 are derived from Van Broekhoven, 1987, Figures 9 to 14 and 21). Taking into consideration the logarithmic scale of the vertical axis, the decrease of the concentration after 1970 is remarkable.

As may be seen from Figure 5 a drastic improvement with regard to the pollution of the Rhine by heavy metals has indeed been achieved after 1970. This change for the better was the result of the joint efforts of the members of the ICPR; since 1982/3 no further decrease has been gained however, though further improvements are badly needed. As far as other harmful pollutants are concerned the situation since 1970 has not improved (e.g. phosphorus, Figure 6) or even worsened (e.g. nitrogen, Figure 6 and chloride, Figure 7) and there are innumerable other contaminants in the Rhine water too! To restore its former good water quality much work still has to be done. An ambitious overall plan to considerably reduce pollution of the Rhine (Rhine action plan) was agreed upon by all ICPR members in October 1987, aiming among other things at a return of salmon and other species of big fish by the year 2000.

The hydrological problems arising from polluted silt will be discussed in view of the situation in the Netherlands. Part of this silt is transported to the North Sea and from there to the Wadden Sea without causing apparent

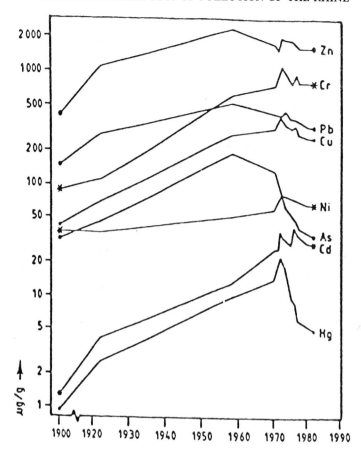

FIGURE 4 *Concentrations of heavy metals in silt sediment, 1900–1985, Nieuwe Merwede the Netherlands (after Salomons, Delft Hydraulics Laboratory, and others)*

hydrological difficulties.[1] The remaining, greater part is deposited in rivers (e.g. the Nieuwe Merwede[2]), in harbours and in freshwater reservoirs. These depositions involve shallowing and subsequently dredging may be required. Formerly the contaminated dredgings were dumped on land or in the North Sea. This option however has become unacceptable: dumping on land makes the area involved utterly unusable (too many huge problems have already arisen from dumping of contaminated dredgings or soil on land in the past),

---

[1]That the pollution of the Rhine and other rivers causes very considerable ecological damage in the North Sea and the Wadden Sea is beyond any doubt.
[2]Local sources, too, may cause such contamination as is the case for the Hollandsche IJssel (N.E. of Rotterdam, Figure 3) with predominantly gravely-polluted deposits over its full length (Anonymous, 1988).

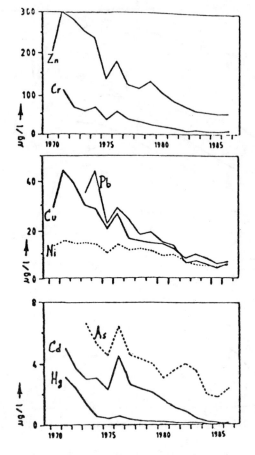

FIGURE 5 *Average concentrations of heavy metals at Lobith*

dumping in the North Sea (a serious threat to its ecology) is inconsistent with the London Dumping Convention (1975).

In a normalized river the banks between the groynes may sometimes offer a temporary solution. For the harbour of Rotterdam, however, from which (apart from other deposits) about 10 million m³ of heavily contaminated dredgings have to be removed each year, specially constructed disposal sites, environmentally safe, are needed (Jansen, 1988). Recently a huge depot of this kind, with a capacitiy of 150 million m³ and situated along the North Sea coast between the two western Rhine outlets (Figure 3), came into use. In the beginning of the next century this 'Slufter' depot (area: 260 ha; costs: DFl 200,000,000) will have reached its capacity; by then silt transported by the Rhine should not be contaminated any more as there are no possibilities of building other big depots. For this there are three reasons:

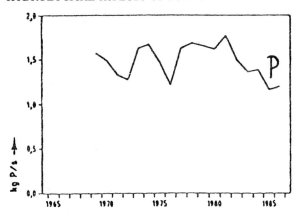

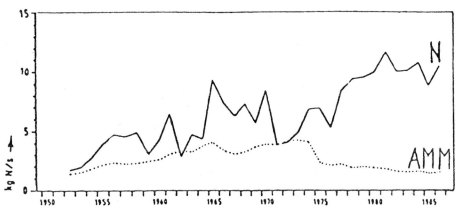

FIGURE 6 *Average loads P and N at Lobith (kg/s)*

(1) it is incompatible with the spirit of the London Dumping Convention;
(2) any such depot should meet great ecological objections;
(3) the costs would be excessive.

As for the freshwater reservoirs (of which the 'western one' receives a certainly not negligible amount of heavily contaminated Meuse silt as well), dredging is not necessary yet, though sooner or later this will be the case. About the most likely further development of these bottoms not much is known for certain; anyhow the situation is not without danger as it is not inconceivable that the toxic substances, adhered to the sediment particles, will get 'mobilized', i.e. dissolve in the water. It goes without saying that such a phenomenon might affect the water quality very badly.

How the problem of contaminated sediments has to be solved is as yet unknown. Some varied possibilities have been suggested but up till now none of them has turned out to be feasible. One bright spot in relation to this

FIGURE 7 *Average loads of chloride at Lobith (kg/s)*

subject appears to be that reports on the contamination of the flood plains of the Rhine in the Netherlands do not qualify them as really dangerously polluted – as is indeed already the case for long stretches of those of the Meuse (for this international river, which rises in France, a commission like the ICPR does not exist).

To avoid misunderstanding: it should be noted that the Netherlands indeed offer a clear example of the various problems produced by contaminated Rhine silt, but this by no means implies that these problems are confined to this country.

River Control and Polluting Accidents

Polluting accidents may greatly spoil the water quality over large reaches of the Rhine. The most notorious is the fire in the Sandoz works near Basle on 1 November 1986, which poisoned the river over a great length; this accident occurred at discharges at Basle some 25% below the mean value. But even accidents outside the Rhine basin may deteriorate the quality of the river water quite badly, as did the disaster at Chernobyl on 26 April 1986, which caused the remaining beta activity at Lobith to reach a value 170 times the mean value of 80 Bq/m^3 on 7 May (Van Broekhoven, 1987). These are

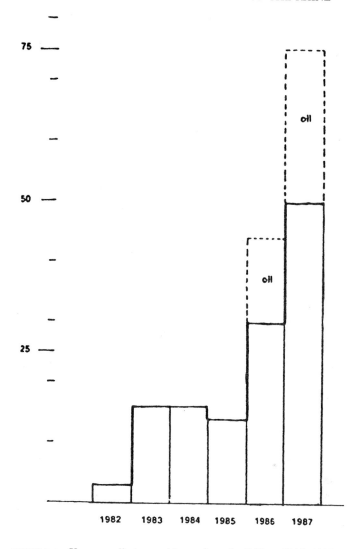

FIGURE 8 *Known polluting accidents along the Rhine, 1982–1987*

evidently extreme cases; less conspicuous accidents are of course not very exceptional, though until rather recently not much was known about them. The graph of known polluting accidents over the period 1982–1987 (Figure 8, derived from Griffioen (1987) and WRK (1988)) therefore does not reflect an increasing dangerousness of industrial processes, but is rather the result of better detection methods and of more reporting of accidents by industries. The latter cause, though not disconnected from the first, is also due to industrial

J. VAN MALDE

top managers becoming increasingly aware of the harmful effects of polluting accidents, both to the river's users and to public relations. It may be clear that the polluting effects of the accidents recorded vary greatly.

Polluting accidents as such obviously do not influence the discharges in the Rhine basin. One might wonder whether in the case of very bad accidents river authorities could interfere with these discharges one way or another in order to reduce as best it can the detrimental consequences of such accidents. On certain stretches and at rather low discharges this would be theoretically possible by manipulating one or more weirs. However, generally speaking, this is not conceivable in practice. For one thing the effects would be quite limited; further, a decision like that would require a balancing of pros and (the very real) cons and thus consultations, whilst often insufficient information is available about the nature and extent of the pollution, etc.

To this state of affairs there may be one exception: the Netherlands. In this country all large rivers and waters are under one authority, Rijkswaterstaat, which eases the procedure of the – still necessary – consultations; also, lying furthest downstream, it has the best chance of adequate information. Finally, the interest of the IJsselmeer and its long replenishment time (two months) constitute a weighty argument for action, should the occasion arise at sufficiently low discharge of the Rhine. If the Nederrijn weirs were then put out of operation the discharge to the IJsselmeer would be reduced (see Table 2), thereby increasing at the same rate the westbound discharge. As by far the majority of the water of this latter discharge would flow into the North Sea within some days, the freshwater reserves would thus be protected as best possible. However, as yet no such policy has been pursued.

An Alarm Model for Polluting Accidents on the Rhine

A pollution calamity on the Rhine is a matter of major concern; yet, the possibilities to control it are very poor indeed. The best hydrologists can do in such a case is to inform authorities as early and accurately as possible on the shape and the propagation of the pollution wave so that they can take their measures swiftly. At their conference in Rotterdam in December 1986, the ministers of the Rhenish riparian states concerned resolved to this end that a joint working group of experts from the ICPR and the CHR were to make a model to compute the propagation and changes in shape of pollution waves. The first version of such an alarm model, covering the Rhine from the Aare mouth east of Basle (Figure 1) downstream, is now ready (Griffioen, 1989). This user-friendly model is to be used on a personal computer; it needs further testing, for which tracer experiments have been and will be done and sensitivity tests are worked out. This will lead to a revised version in about one and a half years' time, which, it is hoped, should also be capable to account for floating pollutants.

REFERENCES

ANONYMOUS (1988) De waterbodem van de Hollandsche IJssel (The bottom of the Hollandsche IJssel). Rijkswaterstaat, Directie Benedenrivieren (in Dutch)

VAN BROEKHOVEN, A.L.M. (1987) De Rijn in Nederland. Toestanden en ontwikkelingen anno 1987 (The Rhine in the Netherlands. State and developments in the year 1987). Internal report Rijkswaterstaat DBW/RIZA, No. 87.061 (in Dutch)

C.H.R./K.H.R. (1978) Le bassin du Rhin. Monographie Hydrologique/Das Rheingebiet. Hydrologische Monographie (The Rhine basin. Hydrological Monograph) (in French and German)

GRIFFIOEN, P.J. (1987) Het Rijn calamiteitenmodel: berekeningen en evaluatie naar aanleiding van Sandoz-affaire (The Rhine calamity model; computation and evaluation on account of the Sandoz affair). H₂O, vol. 20, No. 17 (in Dutch)

GRIFFIOEN, P.J. (1989) Alarm model for the Rhine. Report II. 2 of the C.H.R. published in cooperation with the I.C.P.R. (in German and French)

JANSEN, M.J.D. (1988) Décharges le long du Rhin. (-polluted- Discharges along the Rhine). In: Our river Rhine, a shared responsibility Proceedings International Conference Strasbourg, 3 and 4 March 1988 (in French)

W.R.K. (1988) WRK-jaarrapport 1987 (Annual report of the Watertransport Company Rijn-Kennemerland) (in Dutch)

Hydrological Problems of Dam and Levee Rupture
=

MARIO GALLATI[1] and LUCIO UBERTINI[2]
[1]Pavia University,
[2]Perugia University IRPI/CNR,
Perugia, Italy

Dam Breaking versus Levée Breaking Problem:
Analogies and Differences

SUBMERSION waves originating from the breaking of a dam and of a levée differ in two main characteristics: the spatial display of the wave and the value of potential energy generating the flow field.

In fact the dam-breaking wave usually flows in a narrow valley river bed and it is essentially a unidimensional phenomenon.

Levée breaking produces two-dimensional flows in the close ground plane, or a channel network flow if the direction of the flow is constrained by fixed obstacles as in the case of an urban road network flooding.

As to the hydraulic head, values of the order of 100 m are not unusual for the reservoir head behind a dam, while values of the order of metres are usually reached by the water level of a levée-constrained river. As a consequence, different orders of magnitude characterize the potential energy involved in the two cases. Obviously, many different situations with scales contained between the two above-quoted limiting cases can take place, such as when a monodimensional dam break wave outflows from a narrow valley into a wide plane or when a break is produced in a long earth dam.

Dam Breaking Problems

Physical characteristics and typical space-time scales of the phenomenon

Unfortunately, several well known historical disasters have depicted the effect of the dam breaking submersion wave on the environment and, on the basis of observations and collected data, the causes, water levels and volumes, velocities and time scales have been evaluated.

The water outflowing from a total or partial break in a dam originates a wave characterized by a peak value and a time to peak determined by the time of breach formation. The analysis of several cases has shown that even if it takes time for the formation of the breach, it may be very short so that, 'for the sake of safety', the breaking is often schematized as instantaneous.

The mass of water falling into the valley river bed from a height of maybe 100 m forms a wave advancing downstream, characterized by an abrupt front and followed by a more gradual varying wave-body. Due to the very high bottom slopes typical for mountain valley river beds, supercritical flow conditions occurring at the dam location usually maintained are far downstream. Transitions from supercritical to subcritical flow conditions and vice versa may easily occur due to the effect of the often irregular geometry of the valley cross-section, and increased resistance to the bed modification produced by wave travel.

Analysis of the difficulties and uncertainties connected with simulation of the phenomenon

Usually the simulation of the dam-break submersion wave is required in order to evaluate some significant characteristics of the wave useful to foresee the impact on the environment: the maximum stages reached by the water along the entire valley, the time required by the front to reach particular points where bridges, towns, other structures are located, the speed of the water, correlated with impact effects, erosion, transportation, etc.

The simulation of the whole phenomenon is naturally very difficult and, of course, requires several arbitrary hypotheses in order to quantify esssential characteristics which are impossible to foresee. The results of the simulation must be used to evaluate the order of magnitude of the involved variables and their sensitivity to the governing physical effects. Arbitrary and sometimes unjustified hypotheses are usually introduced to simulate the following aspects:

(1) the formation of the breach;
(2) the water-bed interaction;
(3) the submersion-wave propagation.

(1) The modality of breach formation is quite different for concrete dams and earth dams (Johnson and Illes, 1976; McDonald and Landridge-Monopolis, 1984). In the first case, more than the collapse of the structure itself, the collapse of the rock holding its foundations and landslide waves overlapping the top of the dam are cause of the disaster. Both such geotechnical variables are difficult to predict and quantify. The most common causes of earth dam breaking are connected with water overtopping and foundation soil fluidification piping. The dimensions and

time required for the breach formation have to be assigned on the basis of reasonable hypotheses or even of unreasonable extreme ones like the abrupt dam collapse. A spectrum of simulations based on different excitations will be obtained as a result.

(2) It is practically impossible to foresee the modification of a valley produced by the propagation of the dam break wave and, as it is obvious, this effect is determinant for the evolution of the phenomenon itself. Material is abruptly taken from the valley sides and transported downstream, structures met by the wave front can be destroyed, transported and subsequently accumulated in narrowings of the valley. The obstructions so originated can produce subsequent backwater effects and further collapse.

(3) The problem of simulating the routing of the wave, even through an undeformable bed, can be approached either from physical or numerical basis. Physical simulation is often preferred in the first stages of the phenomenon and when the valley is characterized by abrupt changes of direction, depth and section form, so that form effects, hardly accounted for by a mathematical description, can be properly taken into account.

Mathematical simulation is more suited to the cases where a 'certain' regularity of the bed geometry allows a reasonable schematization of the flow on a monodimensional basis.

In this case, the flow is usually described by the well known open channel flow equations, if possible, written in conservation form (Cunge *et al.*, 1980) taking as flow variable the discharge Q and describing the valley-river bed by means of functions of the water level (h) like, $A(h)$, $B(h)$, $C(h)$, $I(h)$, $I_x(h)$ respectively meaning area, surface width, wetted perimeter, first moment and moment of longitudinal width variation of the cross-section.

The equations are so written in conservation form:

continuity equation
$$\frac{\partial A}{\partial t} + \frac{\partial Q}{\partial x} = 0 \tag{1a}$$

momentum equation
$$\frac{\partial Q}{\partial t} + \frac{\partial M}{\partial x} = gA(i - \mathcal{J}) + gI_x \tag{1b}$$

where: x, t = space and time co-ordinates,

$M = gi + Q^2/A$ = total momentum,

$\mathcal{J} = \dfrac{n^2 Q^2}{R^{4/3} A^2}$ = friction slope (here defined by the Manning formula),

i = bottom slope.

More often, for sake of simplification and reduction of the computational work a different form is used sacrificing the conservative form: the stage variable h is introduced and the following equations are obtained:

$$B \frac{\partial h}{\partial t} + \frac{\partial Q}{\partial x} = 0 \qquad (2a)$$

$$\frac{\partial Q}{\partial t} + \frac{\partial}{\partial x}\left(\frac{Q^2}{A}\right) + gA \frac{\partial h}{\partial x} gA(i - \mathcal{J}) \qquad (2b)$$

The numerically correct solution of the system (Johnson and Illes, 1976) or (Mc Donald and Langridge-Monopolis, 1984), when discontinuities appear, the flow can change from subcritical to supercritical and vice versa during the simulation and dealing moreover with complex geometries, is not an easy task: it requires a careful choice of integration schemes, interpolations, proper position of boundary conditions and, at the end, a very heavy computational work (Cunge *et al.*, 1980).

As the simulation of the phenomenon is affected by the above-quoted inevitable uncertainties and numerical simulation is, in any case, more practical, flexible and less expensive than the physical simulation, even in cases when the basic hypotheses of monodimensionality of the flow is rather uncertain, the numerical simulations is employed at least as a first approach, often followed by a partial or total physical simulation.

The results of the physical model are useful for a correct calibration of the arbitrary parameters used by the numerical model to simulate and quantify complex physical phenomena and interactions often not accounted for by the basic monodimensional assumption.

If the information about the dynamics of the phenomenon is not strictly required, but only an evaluation of the submersion wave maximum levels is required, a strong simplification of the mathematical model can be done for steep rivers by neglecting inertial effects and unbalanced pressure forces. This is justified when the balance of forces reponsible for the dynamic equilibrium, i.e. bottom slope and friction slope, almost balance each other and are of a higher order of magnitude than the other terms of the momentum equation. With this approximation, the kinematic wave model is obtained: it is much easier, safer and cheaper to work out and can be adopted in any case as a first approximation.

The numerical solution of the system (Johnson and Illes, 1976) or (Mc Donald and Langridge-Monopolis, 1984), given its hyperbolic nature and considering the possible subcritical/supercritical transition, could be 'naturally' obtained by the method of characteristic. To the knowledge of the authors, nevertheless, the method has been applied in very few cases with natural geometries (Chervet and Dallères, 1970) as the necessity of spatial interpolation creates heavy errors and requires enormous computer time. The presence of shock waves must moreover be detected and taken into account as an inner moving boundary condition.

Finite difference explicit methods are often preferred also because small time steps are required by computational precision arguments. One of the most popular is, the Lax scheme (Cunge, 1970) for its ease of programming and shock capturing attitude. Leap-frog and diffusive methods have been used, too.

Implicit difference schemes are adopted usefully when it is possible to define in advance the reaches where the flow is either subcritical or supercritical with fixed sections (Frend, 1984). The Preissman-Cunge four-point scheme and Abbot-Jonescu scheme are widely employed for such computations (Cunge et al., 1980).

Current research is focused on the definition of conservative numerical methods are able to deal automatically with discontinuities and subcritical/supercritical flow transitions.

The finite element method used with the Petrov-Galerkin implicit approach (Katopodes, 1984a) or the Taylor-Galerkin (Katopodes and Chien-Ti Wo, 1986) explicit method have shown remarkable proprieties in shock fitting, together with the finite difference Goudonov simplified scheme (Vila, 1987). Such methods have still to be exhaustively tested with irregular geometries.

Beside the numerical schemes problems, practical problems arise particularly for the complex geometry interpolation and the front propagation over dry bed (Cunge et al., 1980).

Levee Breaking Problem

Following the distinction introduced above, we shall deal separately with the two situations of plane submersion wave produced by levée breaking and urban area flooding. In both cases the flow field is spatially two dimensional, even if the flooding of an urban road network can be schematized 'naturally' as an equivalent network of channels connecting accumulation points.

Submersion of a 'natural' plane

Referring to the plane submersion wave, we can distinguish, quite naturally, the part close to the breach, the 'near field', where the flow is dominated by inertial effects and pressure forces and the 'far field' where the flow is influenced essentially by the ground characteristics: slopes, roughness, saturation, etc. The near field flow is rapidly varied and characterized by a small space scale while the far field flow can reasonably be schematized as gradually varied and displaying over a large scale field.

As regards the 'near field', the physical simulation is more suitable than the numerical one especially when dealing with complex geometries and the impact of the submersion wave against obstacles. Both the scale of the problem and the dominating forces are in this case, favourable for the physical simulation. When geometrical boundaries are not too complex, in this case too useful information can be obtained from the numerical simulation. The flow is mathematically described by the well known shallow water equations written in divergence form: the flow variable is either the unit discharge vector or the mean velocity, the geometrical variable is the water height. The correct definition of the boundary conditions according to the subcritical/supercritical regime flow is slightly more complicated than in the one dimensional case and more difficult problems are met in finding the numerical integration schemes well suited for the flow regime. Many questions are still not well defined as classified even at physical level, especially as regards the front propagation over dry bed, and the two-dimensional abrupt front propagation. In fact, even if computer techniques are still at research level, promising results have been already obtained with simplified field geometries (Vila, 1987; Katopodes, 1984b; Van, 1981). Also in this case, the numerical problem is considerably simplified neglecting inertia terms in the shallow water equations for higher water levels simulations but qualitative considerable discrepancies are obtained in the near field. In Figure 1 taken from Braschi and Gallati (1987) a two-dimensional submersion wave is computed both neglecting and taking into account convective terms. It is apparent the effect of convection in determining the characteristic of the 'near' flow field.

As to the large-scale problem, the characteristic time and modalities of the levée breach formation are not such important features as for a dam; moreover, inertial effects can safely be neglected in a simulation aiming to get an estimate of breaking and maximum water levels to draw inundation maps.

Due to these simplifications and to the scale problem the numerical simulation is, in this case, better suited than the physical one (Cunge, 1975; Laura and Wang, 1984).

Practical problems are nevertheless connected with the evaluation and correct definition of the plane roughness that often should have local directional properties and should be assigned as a second order tensor. Laboratory experience has pointed out the importance of the soil saturation in determining the netted front advancement (Van, 1981).

Flooding of an urban area

As for the numerical simulation of an urban area flooding caused by the levée breaking of a river or levée overflow, the flow main direction can be fixed *a priori* on the basis of a simplified urban road network (Braschi *et al.*, 1988). The storage effect can be usefully assigned to the nodes of the network itself assigning a kind of 'urban porosity' to the topological area attaining to the node. The road slopes are often much steeper than channels, the roughness not well defined and form resistance and local losses due to fixed and moving obstacles are quite important and almost unpredictable. The channel flow equations are simplified neglecting inertial effects but upstream or downstream

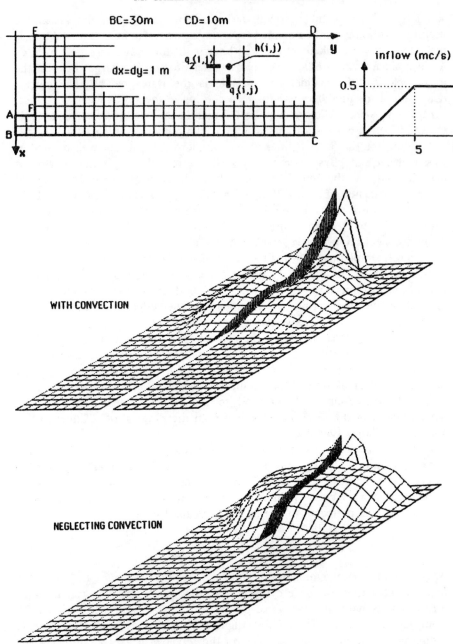

FIGURE 1 *Submersion of a dry plane by a point-like levée breaking wave. Effect of convection. Water level t = 10 s; AB, inflow; DC, outflow; BC, symmetry axis; others, free slip*

184

dependence of the flow must carefully be taken into account to get the correct solution for a complex network. As examples of such computations we show in Figure 2 some computed inundation maps of the town of Sondrio (Italy) based on different reasonable hypotheses of the river Mallero outflow taken from reports (Braschi *et al.*, 1988). The main characteristic of this situation is the steepness of the flow field.

From the same source are also taken some computed maps of an hypothetical flooding of Florence (Italy) due to an outflow of the Arno river: Figure 3. The inflow hydrogram, though extremely simplified, has characteristic scales comparable with those of 1966 flooding. After the equivalance channel network used for the computation, pictures of the growing flooded area at successive times are presented.

Action and Measures Against Dam and Levée Breaking

It is necessary to be fully aware that natural disasters cannot be eliminated. Inundations and droughts will always be present in the world. We must feel bound to know the characteristics, to study the physical processes, to oppose the consequences of these natural phenomena and to live together with them so long as human life and property are protected.

Each country on its own account, and all countries together, must act for the best so that these purposes are realized. Many different sorts of action can be developed, and are developed, at national and international levels. At national level a central political authority (like the Civil Protection Ministry) must co-ordinate other central political authorities (Defence, Ministry, Ministry of the Interior, Environmental Ministry) and different local authorities (region, province, district, municipality, prefecture) in accordance with clear provisions of the law whether for the emergency or for prevention and forecast.

Of course, political authorities must rely on well co-ordinated central and local technical services which are able to follow in real time the evaluation of hydrometeorological events by means of a monitoring network which measures, transmits, elaborates and disseminates the appropriate information needed to take very quick decisions.

At the same time, the technical services must prepare alarm and evacuation plans so that people can behave appropriately once decisions have been made. Natural hazard mitigation needs a synthesis of political co-ordination, technical co-ordination and education.

For the problem of dam breaking, the most important activity is very thorough control during every phase of design, construction and operation; but the control of large dams and small dams can be treated separately. The control of large dams ($h > 10$ m and/or reservoir volume $> 100,000$ m^3) could be assigned to the Dam Office of a Central Ministry such as the Public Works Ministry.

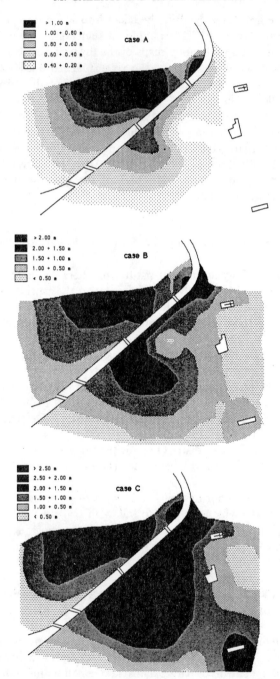

FIGURE 2 *Numerical simulation of Sondrio flooding: maximum height envelope maps relative to different inflow hypotheses*

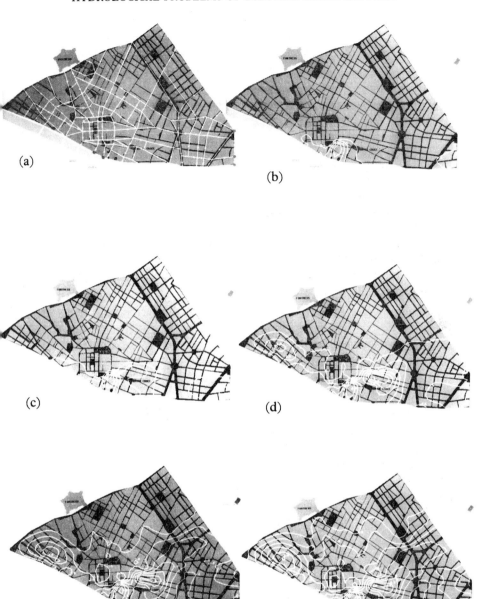

FIGURE 3
Florence flooding: (a) computational network; (b)-(f) water heights computed at different stages of the phenomenon

The control of small dams, whose number (about 20,000 in Italy) is far greater than the number of large dams (about 500 in Italy), could be the task of an office of a local agency of the municipality, province or region.

All the dams must be archived according to a detailed prepared form including project information, engineering data, visual inspection, operation and maintenance procedures, hydraulic and hydrologic characteristics, structural stability. The dams must be inspected periodically by the responsible agency, which must be in a position to ward off possible danger with appropriate and timely assessment and recommendations. Control of the levées can also be done by a central or a local authority, according to the importance of the river. But in this case it is necessary to foresee particular procedures during the floods with continuous, accurate and punctual visual inspection.

At international level it could be hypothesized an initiative to cope with all natural disasters (seismic, vulcanic, flood, landslide, drought, etc.). It can often happen that an area struck by a catastrophe needs technical skills and means which are not present in the state or in the country to which the area belongs. In this case the intervention of an appropriate international unit, which would collaborate with the Government of the affected country would be useful.

As a preliminary idea it is suggested the organization of a Technical-Scientific Emergency Intervention Unit (UFE) be divided into two nuclei:

N1: operative nucleus which must set up field actions in the damaged area;
N2: assistance nucleus which must co-operate with N1 to furnish data and information on different subjects involved with the disaster (people, government, local, central and international agencies).

The nuclei could be organized into three different sectors: Technical, Informative, Recording.

The activities of three sectors could be:

(1) Technical
(a) To build works and evacuation plans of first intervention (bridges, roads, canals, physical and mathematical models, etc.);
(b) to furnish detailed designs for immediate or short-term interventions;
(c) to furnish preliminary designs for medium or long-term interventions;
(d) to build monitoring systems appropriate to the occurrence of hydrogeological disasters;
(e) to furnish assistance for the control of monitoring system and the adequate software of forecast models.

(2) Informative
(a) To develop a survey activity of hydrogeological conditions of the event, to evaluate its relative and absolute gravity and its future evolution. The collected data are sent to the government authority, which can furnish an

impeccable and immediate supply of information to different parties involved (mass-media, people, etc.);

(b) to take a census of property and infrastructural damages to furnish to the government authority the support for administrative refund procedures.

(3) Recording

(a) To build and manage the database of the event for immediate statistical analysis;
(b) to compare the hydrological and technical characteristics of the event with other antecedent events in the same area and in the other areas.

As a consideration, we would like to outline the specific contributions the scientific community could make. The experience gained in Italy over the last four years has shown some results of undoubted interest.

In 1984 the Ministry of Civil Protection, the Ministry of Scientific Research and Ministry of Public Works issued a law which established the national group for hydrogeological disaster preservation.

The group co-ordinates the research and operative activity of 97 operative units (Institutes of Universities and of the National Research Council) with the following tasks:

(a) to promote, develop and co-ordinate multidisciplinary researches on floods and landslides;
(b) to furnish scientific advice to ministries, regional and local authorities, with particular emphasis on civil protection problems;
(c) to assure the co-ordination of scientific intervention on the occasion of floods and landslides phenomena;
(d) to formulate proposals for specific research programs;
(e) to formulate proposals for rules and actions regarding prevention and forecasting;
(f) to promote exchange of international scientific experiences on floods and landslides.

The programme of the group is subdivided into four lines of research:

(1) forecasting, prevention and monitoring of hydrological extremes;
(2) forecasting and prevention of landslide events of high risk;
(3) evaluation of hydrogeological risk and zoning; strategies for the mitigation of the effects of extreme events.

(4) Evaluation of vulnerability of aquifers

When an emergency happens the Ministry of Civil Protection ask for the director of the group the identification of the OU able to cope with the emergency.

The most appropriate OUs are identified on the basis of scientific ability and knowledge of the damaged area.

In the last three years more than 200 interventions have been made by the group. During the Valtellina event, 30 researchers of the group worked for more than three months. In that case a big landslide was forecast two days in advance and it was possible to evacuate two villages which were later completely destroyed.

In accordance with the resolutions of US-ASIA conference on 'Engineering for the mitigation of natural hazards damage' (Bangkok, 14–18 December 1987) and of the International Workshop on Natural Disasters in European-Mediterranean countries (Perugia, 27 June to 1 July 1988), the priority of research activity must be assigned to flood prediction from failure of dams and flow retaining structures.

In recent years considerable attention has been given to this problem. Most of the work done stems from dam safety investigations carried out to determine the safety of the dam.

Likewise, some research has been done to examine modes of dam failure, to model the breach process mathematically and to estimate the resulting flood hydrographs.

The general goal of this research is to improve existing procedures and to develop new procedures for estimating the flood and sediment hydrograph resulting from the failure of dams and levées.

The specific objectives are:

(1) To identify and test an appropriate model to determine the progression of the dam or levée breach resulting from different triggering mechanisms overtopping and sudden collapse of part of the structure. Various boundary conditions must be considered as well as different designs of the structures.
(2) To develop and test mathematical and physical models for estimating the flood hydrographs from dam and levée breaches. A number of factors need to be investigated more accurately. These factors include: the geometry of the embankments, structural materials, initial breach configuration, boundary conditions, effect of sediment and flow conditions and geometry of the channel.

We would stress, however, that the scientific community, though conscious of its own limits has been, is, and will continue to be available to furnish its own contribution, because of its appreciation that work on these problems is very satisfactory not only from the scientific, but also from the social point of view.

REFERENCES

BRASCHI, G. and GALLATI, M. (1987) Simulation of flooding in a two dimensional field. Proceedings of 18th Modeling and Simulation Conference, Pittsburg

BRASCHI, G., GALLATI, M. and NATALE, L. (1988) La simulazione delle inondazioni in ambiente urbano. Int. Rept. no 152 of Dept. of Hydraulic and Environmental Engineering, University of Pavia, Italy

CHERVET, A. and DALLÈRES, P. (1970) Calcul de l'onde de submersion consecutive à la rupture d'un barrage. *Schweiz Banz* **88**: no 19

CUNGE, A. (1970) Calcul de propagation des ondes de rupture de barrage. *La Houille Blanch* **1**: 15–23

CUNGE, J.A. (1975) Two-dimensional modelling of flood plains. *Unsteady Flow in Open Channels*, Water Reserach Publications, Fort Collins

CUNGE, J.A., HOLLY, F.M. and VERWEY, A. (1980) *Practical aspects of computational river hydraulics*. Pitman Advanced Publishing Company, London

JOHNSON, F.A. and ILLES, P. (1976) A classification of dam failures. *Water Power Dam Constr* **28**: 43–45.

KATOPODES, N.D. (1984a) A dissipative Galerkin scheme for open channel flows. *Proc ASCE J Hydr Eng* **110**: no 4

KATOPODES, N.D. (1984b) Two dimensional surges and shocks in open channels. *Proc ASCE J Hydr Eng* **110**: no 6

KATOPODES, N.D. and CHIEN-TAI WO (1986) Explicit computation of discontinuous channel flows. *Proc ASCE J Hydr Eng* **112**: no 6

LAURA, R.A. and WANG, J.D. (1984) Two dimensional flood routing on steep slopes. *Proc ASCE J Hydr Eng* **110**: no 8

Mc DONALD, T.C. and LANGRIDGE-MONOPOLIS, J. (1984) Breaching characteristics of dam failures. *Proc ASCE J Hydr Eng* **110**: 567–586

VAN, M. (1981) *Modèle bidimensionnelle d'onde de rupture de barrage – Comparaison mesures-calcul pour un cas schematique, E43/81–08.* Laboratoire National d'Hydraulique, EDS, France

VILA, J.P. (1987) Schemas numeriques en hydraulique des écoulements avec discontinuités. *Proc IAHR, Lausanne*, 120–125

National Weather Service Models to Forecast Dam-Breach Floods

—

D.L. FREAD

Hydrologic Research Laboratory, Office of Hydrology,
National Weather Service, NOAA, Silver Spring, Maryland, USA

Introduction

DAMS provide society with essential benefits such as water supply, flood control, recreation, hydropower, and irrigation. However, catastrophic flooding occurs when a dam fails, and the impounded water escapes through the breach into the downstream valley. Usually, the magnitude of the flow greatly exceeds all previous floods, and warning time is reduced. The National Weather Service (NWS) has the responsibility of warning the downstream populace if a dam fails. In the United States, there are some 70,000 dams (height > 25 ft). Many of these are more than 30 years old and present increased hazard potential due to downstream development and increased risk of failure due to structural deterioration or inadequate spillway capacity. The NWS has developed three models to aid flash flood hydrologists in forecasting dam breach flooding. These models (DAMBRK, SMPDBK, and BREACH) are coded in Fortran, are available at a nominal cost, and may be executed on mainframe, mini, or microcomputers by hydrologists/engineers concerned with hazard mitigation planning or spillway design.

DAMBRK, initially developed in 1977, has been adopted for use in the United States by most federal and state agencies concerned with dam safety and design. Also, DAMBRK is being used in several countries within the Americas and elsewhere around the world by governmental agencies, power companies, and private engineering consultants. Research has been ongoing in developing improvements in the DAMBRK model allowing it to have an increasing range of application and improved user friendly characteristics. The model can be used as part of a real-time hazard warning system or for developing hazard mitigation plans. Also, DAMBRK can be used for routing any specified hydrograph through reservoirs, rivers, canals, or estuaries as part of general engineering studies of waterways. Recently, DAMBRK has been expanded to route mud/debris flow hydrographs. SMPDBK, developed in 1982–83, is a much simplified technique for predicting dam-breach flooding. It has received considerable use in the United States when available time and resources are too limited for the use of DAMBRK and the attendant reduction in accuracy is judged acceptable. BREACH, developed in 1984–85, predicts only the breach hydrograph and the breach size and its time of formation due to overtopping or piping of earthen-rockfill dams. It has thus far received rather limited application for selecting the breach parameters required by DAMBRK and SMPDBK. BREACH has promise to reduce the uncertainty associated with selecting breach parameters; however it should be used with judgement and caution until it receives further verification to determine its extent of applicability and reliability. This paper presents a brief description and an application of each model.

DAMBRK

The DAMBRK model (Fread 1977, 1984a, 1985, 1988) is representative of the current state-of-the-art in understanding dam failures and the utilization of hydrodynamic theory to predict the dam-break wave formation and downstream progression. The model has wide applicability; it can function with various levels of input data ranging from rough estimates to complete data specification; the required data is readily accessible; and it is economically feasible to use, i.e. it requires minimal computational effort on mainframe computing facilities and can be used with microcomputers. DAMBRK is used to develop the outflow hydrograph from a dam and hydraulically route the flood through the downstream valley. The governing equations of the model are the complete one-dimensional Saint-Venant equations of unsteady flow which are coupled with internal boundary equations representing the rapidly varied flow through structures such as dams and bridges/embankments which may develop a time-dependent breach. Also, appropriate external boundary equations at the upstream and downstream ends of the routing reach are utilized. The system of equations is solved by a nonlinear weighted four-point implicit finite difference method. The flow may be either subcritical or supercritical with fluid properties obeying either Newtonian or non-Newtonian plastic principles. The hydrograph to be routed may be specified as an input time series or it can be developed by the model using specified breach parameters (size, shape, time of development). The possible presence of downstream dams which may be breached by the flood, bridge/embankment flow constrictions, tributary inflows, river sinuosity, levées located along the downstream river, and tidal effects are each properly considered during the downstream propagation of the flood. DAMBRK also may be used to route mud and debris flows or rainfall/snowmelt floods using specified upstream hydrographs. High water profiles along the valley, flood arrival times, and hydrographs at user selected locations are standard model output.

Saint-Venant equations

A modified and expanded form of the original Saint-Venant equations (Delong, 1985; Fread, 1984a) consist of a conservation of mass equation, i.e.

$$\frac{\partial Q}{\partial X} + \frac{\partial s_c (A + A_0)}{\partial t} - q = 0 \qquad (1)$$

and a conservation of momentum equation, i.e.

$$\frac{\partial (s_m Q)}{\partial t} + \frac{\partial (Q^2/A)}{\partial x} + gA(\frac{\partial h}{\partial x} + S_f + S_e + S_i) + L = 0 \qquad (2)$$

where h is the water elevation, A is the active cross-sectional area of flow, A_0 is the inactive (off-channel storage) cross-sectional area, s is a sinuosity factor (Delong, 1985) which varies with h, x is the longitudinal distance along the channel (valley), t is the time, q is the lateral inflow or outflow per linear distance along the channel (inflow is positive and outflow is negative in sign), g is the acceleration due to gravity, S_f is the boundary friction slope, S_e is the expansion–contraction slope, and S_i is the additional friction slope associated with internal viscous dissipation of non-Newtonian fluids such as mud/debris flows. The boundary friction slope is evaluated from Manning's equation for uniform, steady flow, i.e.

$$S_f = \frac{n^2 \, |Q| \, Q}{2.21 \, A^2 \, R^{4/3}} = |Q| \, Q/K^2 \qquad (3)$$

in which n is the Manning coefficient of frictional resistance, R is the hydraulic radius, and K is the channel conveyance factor. The term (S_e) is

$$S_e = \frac{k \, \Delta(Q/A)^2}{2g \, \Delta x} \qquad (4)$$

in which k is the expansion-contraction varying from 0.0 to $\pm$ 1.0 (+ if contraction, $-$ if expansion), and $\Delta(Q/A)^2$ is the difference in the term $(Q/A)^2$ at two adjacent cross-sections separated by a distance Δx. L is the momentum effect of lateral flow assumed herein to enter or exit perpendicular to the direction of the main flow. This term has the following form: (1) lateral inflow, $L = 0$; (2) seepage lateral outflow, $L = -0.5 \, qQ/A$; and (3) bulk lateral outflow, $L = -qQ/A$. The term (S_i) can be significant only when the fluid is non-Newtonian. It is evaluated for any non-Newtonian flow as follows:

$$S_i = \frac{\varkappa}{\gamma} \left[\frac{(b + 2)Q}{AD^{b+1}} + \frac{(b + 2) \, (\tau_0/\varkappa)^b}{2D^b} \right]^{1/b} \qquad (5)$$

in which γ is the fluid's unit weight, τ_0 is the fluid's yield strength, D is the

194

hydraulic depth (ratio of wetted area to top width), $b = 1/m$ where m is the power of the power function that fits the fluid's stress-strain properties, and $\varkappa$ is the apparent viscosity or scale factor of the power function.

Equations (1) and (2), which are nonlinear partial differential equations, must be solved by numerical techniques. An implicit four-point finite difference technique is used to obtain the solution. This particular technique is used for its computational efficiency, flexibility, and convenience in the application of the equations to flow in complex channels existing in nature. In essence, the technique determines the unknown quantities (Q and h at all specified cross-sections along the downstream channel/valley) at various times into the future; the solution is advanced from one time to a future time by a finite time interval (time step) of magnitude Δt. The flow equations are expressed in finite difference form for all cross-sections along the valley and then solved simultaneously for the unknowns (Q and h) at each cross-section. Due to the nonlinearity of the partial differential equations and their finite difference representations, the solution is iterative and a highly efficient quadratic iterative technique known as the Newton-Raphson method is used. Convergence of the iterative technique is attained when the difference between successive iterative solutions for each unknown is less than a relatively small prescribed tolerance. Usually, one to three iterations at each time step are sufficient for convergence to be attained for each unknown at all cross-sections. A more complete description of the solution technique may be found elsewhere (Fread 1984, 1985).

Internal boundaries

A dam is considered an internal boundary which is defined as a short Δx reach between sections i and $i + 1$ in which the flow is governed by the following two equations rather than Equations (1) and (2):

$$Q_i = Q_{i + 1} \qquad (6)$$

$$Q_i = Q_s + Q_b \qquad (7)$$

in which Q_s and Q_b are the spillway and breach flow, respectively. In this way, the flows Q_i and $Q_{i + 1}$ and the elevations h_i and $h_{i + 1}$ are in balance with the other flows and elevations occurring simultaneously throughout the entire flow system which may consist of additional dams which are treated as additional internal boundary conditions via Equations (6) and (7). In fact, DAMBRK can simulate the progression of a dam-break flood through an unlimited number of reservoirs located sequentially along the valley. The downstream dams may also breach if they are sufficiently overtopped. The spillway flow (Q_s) is computed from the following expression:

$$Q_s = c_s L_s(h - h_s)^{1.5} + c_g A_g(h - h_g)^{0.5} + c_d L_d(h - h_d)^{1.5} + Q_t \qquad (8)$$

in which c_s is the uncontrolled spillway discharge coefficient, h_s is the uncontrolled spillway crest, c_g is the gated spillway discharge coefficient, h_g is the centre-line elevation of the gated spillway, c_d is the discharge coefficient for flow over the crest of the dam, L_s is the spillway length, A_g is the gate flow area, L_d is the length of the dam crest less L_s, and Q_t is a constant outflow term which is head independent. The uncontrolled spillway flow or the gated spillway flow can also be represented as a table of head-discharge values. The gate flow may also be specified as a function of time. The breach outflow (Q_b) is computed as broad-crested weir flow, i.e.

$$Q_b = c_v k_s [3.1 \, b_i \, (h - h_b)^{1.5} + 2.45 \, z \, (h - h_b)^{2.5}] \qquad (9)$$

in which c_v is a small correction for velocity of approach, b_i is the instantaneous breach bottom width, h is the elevation of the water surface just upstream of the structure, h_b is the elevation of the breach bottom which is assumed to be a linear function of the breach formation time, z is the side slope of the breach, and k_s is the submergence correction due to downstream tailwater elevation (h_t), i.e.

$$k_s = 1.0 - 27.8 \left[\frac{h_t - h_b}{h - h_b} - 0.67 \right]^3 \qquad (10)$$

If the breach is formed by piping, Equation (9) is replaced by an orifice equation:

$$Q_b = 4.8 \, A_p (h - h_f)^{1/2} \qquad (11)$$

where:

$$A_p = [2b_i + 4z(h_f - h_b)](h_f - h_b) \qquad (12)$$

in which h_f is the specified centre-line elevation of the pipe.

Highway/railway bridges and their associated earthen embankments which are located at points downstream of a dam may also be treated as internal boundary conditions. Equations (6) and (7) are used at each bridge; the term Q_s in Equation (7) is computed by the following expression:

$$Q_s = C\sqrt{g} \, A_{i + 1}(h_i - h_{i + 1})^{0.5} + C_d k_s (h - h_c)^{1.5} \qquad (13)$$

in which C is a coefficient of bridge flow, C_d is the coefficient of flow over the crest of the road embankment, h_c is the crest elevation of the embankment, and k_s is similar to Equation (10). A breach of the embankment is treated the same as with dams.

Breach

The breach is the opening formed in the dam as it fails. Earthen dams which exceedingly outnumber all other types of dams do not tend to completely fail, nor do they fail instantaneously. The fully formed breach in earthen dams tends to have an average width (b) in the range ($h_d < b < 5h_d$) where h_d is the height of the dam. The middle portion of this range for b is supported by the summary report of Johnson and Illes (1976). Breach widths for earthen dams are therefore usually much less than the total length of the dam as measured across the valley. Also, the breach requires a finite interval of time (τ) for its formation through erosion of the dam materials by the escaping water. Total time of failure may be in the range of a few minutes to a few hours, depending on the height of the dam, the type of materials used in construction, the extent of compaction of the materials, the magnitude and duration of the overtopping flow of the escaping water. Piping failures occur when initial breach formation takes place at some point below the top of the dam due to erosion of an internal channel through the dam by escaping water. As the erosion proceeds, a larger and larger opening is formed; this is eventually hastened by caving-in of the top portion of the dam. Concrete gravity dams also tend to have a partial breach as one or more monolith sections formed during the construction of the dam are forced apart by the escaping water. The time for breach formation is in the range of a few minutes. Poorly constructed earthen dams and coal-waste slag piles which impound water tend to fail within a few minutes, and have average breach widths in the upper range or even greater than those for the earthen dams mentioned above.

Recently, Froelich (1987), using the properties of 43 breached dams ranging in height from 15 to 285 ft, with all but six between 15 and 200 ft, presented statistically derived predictors for b and τ. From Froelich's work, the following predictive equations can be obtained:

$$\bar{b} = 9.5k_o(V_r h_d)^{0.25} \tag{14}$$

$$\tau = 0.6(V_r/h_d^2)^{0.50} \tag{15}$$

in which $\bar{b}$ is the average breach width (ft), τ is time of failure (hours), $\varkappa_0 = 0.7$ for piping and 1.0 for overtopping failures modes, V_r is volume (acre-ft), and h_d is height (ft) of water over the final breach bottom which is usually about the height of the dam. Standard error of estimates for $\bar{b}$ was $\pm$ 94 ft, which is an average error of $\pm$ 54% of $\bar{b}$, and the standard error of estimate for τ was $\pm$ 0.9 hour, which is an average error of $\pm$ 70% of τ.

In DAMBRK, the failure time (τ) and the size and shape of the breach are selected as input parameters similar to the approach used by Fread and Harbaugh (1973). The shape is specified by a parameter (z) identifying the side slope of the breach, i.e. one vertical: z horizontal. Rectangular, triangular, or trapezoidal shapes may be specified in this way. For example, $z > 0$ and $b > 0$ produces a trapezoidal shape. The final breach size is controlled by the z parameter and another parameter (b) which is the terminal width of the bottom of the breach. The model assumes the breach bottom width starts at a point and enlarges at a linear rate over the failure time (τ) until the terminal width is attained and the breach bottom has eroded to the elevation h_{bm} which is usually, but not necessarily, the bottom of the reservoir or outlet channel

197

bottom. During the simulation of a dam failure, the actual breach formation commences when the reservoir water surface elevation (h) exceeds a specified value, h_f. This feature permits the simulation of an overtopping of a dam in which the breach does not form until a sufficient amount of water is flowing over the crest of the dam. A piping failure is simulated when h_f is specified less than the height of the dam, h_d.

External boundaries

The upstream boundary (a known relationship between flow and depth or time) is usually $Q_1 = QI(t)$. If the water surface of the most upstream reservoir is assumed to remain level as it varies with time, then the following boundary equation is used:

$$Q_1 = QI(t) - 0.5 \, \bar{S}_a \, 43560. \; \Delta h / \Delta t \qquad (16)$$

in which Q_1 is the discharge at the upstream most section (the upstream face of the dam), $QI(t)$ is the specified inflow to the reservoir, $\bar{S}_a$ is the average surface area (acre-ft) of the reservoir during the Δt time interval, and Δh is the change in reservoir elevation during the time step. If the flow is supercritical throughout the routing reach, two boundary equations are used at the upstream section, i.e. $Q_1 = QI(t)$ and $Q_1 = KS^{0.5}$ in which K is the channel conveyance and S is the instantaneous water surface slope.

For subcritical flows, a known relationship between flow and depth or time must be specified for the most downstream section. The downstream boundary is often a rating table of discharge, associated with a particular depth. It may also be known water elevation variation with time such as a large tidal bay or lake. If the flow is supercritical, no downstream boundary is required since downstream flow disturbances do not progress upstream.

Subcritical/supercritical algorithm

This optional algorithm (Fread 1983, 1985, 1988) automatically subdivides the total routing reach into subreaches in which only subcritical (Sub) or supercritical (Sup) flow occurs. The transition locations where the flow changes from Sub to Sup or vice versa are treated as external boundary conditions. This avoids the application of the Saint-Venant equations to the critical flow transitions. At each time step, the solution commences with the most upstream subreach and proceeds subreach by subreach in the downstream direction. The upstream boundary (UB) and downstream boundary (DB) are automatically selected as follows: (1) when the most upstream subreach is Sub, the UB is the specified discharge hydrograph and the DB is the critical flow equation since the next downstream subreach is Sup; (2) when the most upstream subreach is Sup, the UB is the specified hydrograph and a loop-rating quite similar to that previously described as an external boundary condition, and no DB is required since flow disturbances created downstream of the Sup reach cannot propagate upstream into this subreach; (3) when an inner subreach is Sup, its two UB conditions are the discharge just computed at the DB of the adjacent upstream subreach and the computed critical water surface elevation at the same DB; (4) when an inner subreach is Sub, its UB is the discharge just computed at the most downstream section of the adjacent upstream Sup subreach and the DB is the critical flow equation. Hydraulic jumps are allowed to move either upstream or downstream prior to advancing to another computational time step; this is accomplished by comparing computed sequent elevations (h_s) with computed backwater elevations (h) at each section in the vicinity of the hydraulic jump. The jump is moved section by section upstream until $h_s > h$ or moved downstream until $h > h_s$. The Froude number (Fr = $\sqrt{Q/(gDA^2)}$) is used to determine if the flow at a particular section is Sub or Sup, i.e. if $Fr < 1$ the flow is Sub and if $Fr \geqq 1$ the flow is Sup. The Sub/Sup algorithm increases the computational requirements by approximately 20%.

Lateral flows

Unsteady flows associated with tributaries upstream or downstream of a dam can be added to the unsteady flow resulting from the dam failure. This is accomplished via the term q in Equation (1). The tributary flow is distributed along a single Δx reach. Backwater effects of the dam-break flow on the tributary flow are ignored, and the tributary flow is assumed to enter perpendicular to the dam-break flow. Outflows are assigned negative values. Outflows which occur as broad-crested weir flow over a levée or natural crest may be simulated. The crest elevation, discharge coefficient, and location along the river valley must be specified. The head is computed as the average water surface elevation, along the crest length, less the crest elevation.

Floodplain compartments

The DAMBRK model can simulate the exchange of flow between the river and floodplain compartments. The floodplain compartments are formed by one or two levées which run parallel to the river on either or both sides of the river, and other levées or road embankments which run perpendicular to the river. Flow transfer between a floodplain compartment and the river is assumed to occur along adjacent Δx reaches and is controlled by broad-crested weir flow with submergence correction. Flow can be either away from the river or into the river, depending on the relative water surface elevations of the river and the floodplain compartment. The river elevations are computed via Equations (1) and (2), and the floodplain water surface elevations are computed by a simple storage routing relation, i.e.

$$V_l^t = V_l^{t-\Delta t} + (I^t - O^t)\,\Delta t/43560 \tag{17}$$

in which V_l is the volume (acre-ft) in the floodplain compartment at time t or $t - \Delta t$ referenced to the water elevation, I is the inflow from the river or adjacent floodplain compartments, and O is the outflow from the floodplain compartment to the river and/or to adjacent floodplain compartments. Flow transfer between adjacent floodplain compartments is controlled by broad-crested weir flow with submergence correction. The outflow from a floodplain compartment may also include that from one or more pumps associated with each floodplain compartment. Each pump has a specified discharge-head relation given in tubular form along with specified start-up and shut-off operating elevations. The pumps discharge to the river.

Landslide waves

The capability to generate a wave produced by a landslide, which rushes into a reservoir, is provided within DAMBRK. The volume of the landslide mass its porosity, and time interval over which the landslide occurs, are input to the model. Within the model, the landslide mass is deposited within the reservoir in layers during small computational time steps, and simultaneously the original dimensions of the reservoir are reduced accordingly. The time rate of reduction in the reservoir cross-sectional area creates the wave during the solution of the unsteady flow Equations (1) and (2), applied to the cross-sections describing the reservoir characteristics. The wave may have sufficient amplitude to overtop the dam and precipitate a failure of the dam, or the wave by itself may be large enough to cause catastrophic flooding downstream of the dam without resulting in the failure of the dam as perhaps in the case of a concrete dam.

Automatic selection of Δx and Δt computational steps

The computational distance steps (Δx) and/or the computational time steps may be selected by the user. However, there is an option to let the program automatically select either or both steps. The distance steps are automatically selected to obey the most restrictive of the following criteria:

$$\Delta x = L/(1 + 2 \mid A_i - A_{i+1} \mid /A) \qquad (18)$$

$$\Delta x < c\, \Delta t \qquad (19)$$

The first criteria is derived from Samuels' theoretical work (Samuels, 1985) which indicated the four-point implicit difference scheme is limited to changes in cross-sectional area (expansions or contractions) within a computational distance step to: $0.635 < A_{i+1}/A_i < 1.576$. In Equation (18), L is the distance between specified sections i and $i + 1$, A is A_{i+1} for contractions and A_i for expansions, and Δx is the necessary computational distance step(s), subdividing L such that the inequality criterion of Samuels is satisfied. Equation (19) also restricts the Δx according to c (the bulk wave speed) times the computational distance step. Further restrictions on Δx may be imposed in the vicinity of sharp breaks in the channel bottom slope. The computational time step is automatically selected according to the following criterion:

$$\Delta t = T_p/20$$

in which T_p is the time of rise of the most abrupt wave existing in the flow system at the moment when the time step is to be used. Thus T_p may represent the inflow hydrograph or the breach initiated hydrograph in which T_p is replaced by τ, the time of failure. If the dam has not yet breached, a larger time step is used since $T_p \gg \tau$; yet when the dam begins to breach, a smaller time step is then used henceforth.

Model testing

The DAMBRK model has been tested on several historical floods due to breached dams to determine its ability to reconstitute observed downstream peak stages, discharges, and travel times. Among the floods that have been used in the testing are: 1976 Teton Dam, 1972 Buffalo Creek Coal-Waste Dam, 1889 Johnstown Dam, 1977 Toccoa (Kelly Barnes) Dam, and the 1977 Laurel Run Dam floods. However, only the Teton and Buffalo Creek peak flow profiles are presented herein.

The Teton Dam, a 300 ft (91.4 m) high earthen dam with 230,000 acre-ft (282,900,000 m³) of stored water and maximum 262.5 ft (80 m) water depth, failed on 5 June 1976, killing 11 people making 25,000 homeless and inflicting about $400 million in damages to the downstream Teton-Snake River Valley. Data from a Geological Survey Report (Ray et al., 1976) provided observations on the approximate development of the breach, description of the reservoir storage, downstream cross-sections and estimates of Manning's n approximately every five miles, estimated peak discharge measurements of four sites, flood peak travel times, and flood peak elevations. The critical breach parameters were $\tau = 1.43$ hours, $b = 80$ ft (24.4 m), and $z = 1.04$. The computed peak flow profile along the downstream valley is shown in Figure 1. Variations between computed and observed values are less than 5%.

The Buffalo Creek 'coal waste' dam, a 44 ft (13.4 m) high tailings dam with 400 acre-ft (492,000 m³) of storage failed on 26 February 1972, resulting in 118 lives lost and over $50 million in property damage. Flood observations (Davies et al., 1975) along with the computed flood peak profile extending about 16 miles downstream are shown in Figure 2. Critical breach parameters were $\tau = 0.08$ hours, $b = 170$ ft (49.2 m), and $z = 2.6$. Comparison of computed and observed flows indicate an average difference of about 9%.

SMPDBK

The SMPDBK model, as described in detail by Wetmore and Fread (1984), is a simple model for predicting the characteristics of the floodwave peak

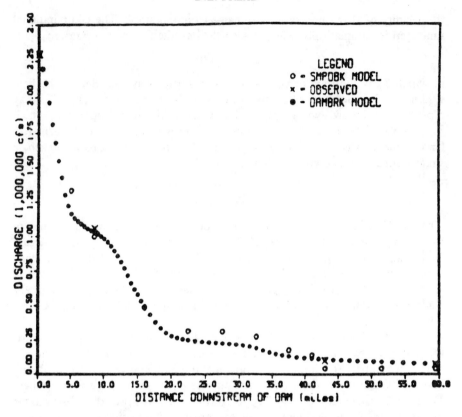

FIGURE 1 *Profile of peak discharge downstream of Teton*

produced by a breached dam. It will, with minimal computational resources
(hand-held calculators, microcomputers), determine the peak flow, depth, and
time of occurrence at selected locations downstream of a breach dam.
SMPDBK first computes the peak outflow at the dam, based on the reservoir
size and the temporal and geometrical description of the breach. The
computed flood-wave and channel properties are used in conjunction with
routing curves to determine how the peak flow will be diminished as it moves
downstream. Based on this predicted floodwave reduction, the model
computes the peak flows at specified downstream points. The model then
computes the depth reached by the peak flow based on the channel geometry,
slope, and roughness at these downstream points. The model also computes
the time required for the peak to reach each forecast point and, if a flood
depth is entered for the point, the time at which that depth is reached, as well
as when the floodwave recedes below that depth, thus providing a time frame
for evacuation and fortification on which a preparedness plan may be based.
The SMPDBK model neglects backwater effects created by downstream dams
or bridge embankments, the presence of which can substantially reduce the

202

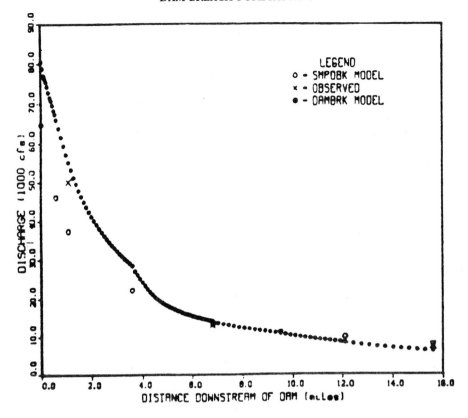

FIGURE 2 *Profile of peak discharge downstream of Buffalo Creek*

model's accuracy. However, its speed and ease of use, together with its small computational requirements, make it an attractive tool for use in cases where limited time and resources preclude the use of the DAMBRK model. In such instances, planners, designers, emergency managers, and consulting engineers responsible for predicting the potential effects of a dam failure may employ the model where backwater effects are not significant.

The SMPDBK model retains the critical deterministic components of the DAMBRK model while eliminating the need for extensive numerical computations. It accomplishes this by approximating the downstream channel/valley as a prism, concerning itself with only the peak flows, stages, and travel times, neglecting the effects of backwater from downstream bridges and dams, and utilizing dimensionless peak-flow routing graphs developed by using the DAMBRK model. The applicability of the SMPDBK model is enhanced with its user friendly interactive input and option for minimal data requirements. The peak flow at the dam may be computed with only four readily accessible data values, and the downstream channel/valley may be

203

defined with a single average cross section, although prediction accuracy increases with the number of specified cross-sections.

Breach outflow

The model uses a single equation to determine the maximum breach outflow and the user is required to supply the values of four variables for this equation. These variables are: (1) the surface area (A_s, acres) of the reservoir; (2) the depth (H, ft) to which the breach cuts; (3) the time (t_f, minutes) required for breach formation; and (4) the final width (B_r, ft) of the breach. These parameters are substituted into an analytically derived time-dependent broad-crested weir flow equation to yield the maximum breach outflow (Q_{bmax}) in cfs, i.e.

$$Q_{bmax} = Q_o + 3.1\ B_r \left(\frac{C}{t_{f/60} + C/\sqrt{H}}\right)^3 \tag{20}$$

where:

$$C = \frac{23.4\ A_s}{B_r} \tag{21}$$

and Q_o is the spillway flow and overtopping crest flow which is estimated to occur simultaneously with the peak breach outflow.

Once the maximum outflow at the dam has been computed, the depth of flow produced by this discharge may be determined based on the geometry of the channel immediately downstream of the dam, the Manning 'n' (roughness coefficient) of the channel and the slope of the downstream channel. This depth is then compared to the depth of water in the reservoir to find whether it is necessary to include a submergence correction factor for tailwater effects on the breach outflow, i.e. to determine if the water downstream is restricting the free flow through the breach. This comparison and (if necessary) correction allows the model to provide the most accurate prediction of maximum breach outflow which properly accounts for the effects of tailwater depth downstream of the dam. The submergence correction is computed from Equation (10) and must be applied iteratively since the outflow produces the tailwater depth which determines the submergence factor which affects the outflow.

Peak flow routing

The peak discharge is routed to downstream points of interest through the channel/valley described by selected cross-sections defined by tables of widths and associated elevations. The routing reach from the dam to the point of interest is approximated as a prismatic channel by defining a single cross-section (an average section that incorporates the geometric properties of all intervening sections via a distance weighting technique). This prismatic

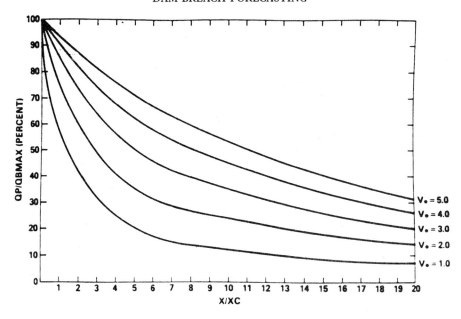

FIGURE 3 *Routing curves for SMPDBK model for Froude no = 0.25*

representation of the channel allows easy calculation of flow area and volume in the downstream channel which is required to accurately predict the amount of peak flow attenuation. The peak flow at the dam computed from Equation (20) is routed downstream using the dimensionless routing curves (see Figure 3). These curves were developed from numerous executions of the NWS DAMBRK model and they are grouped into families based on the Froude number associated with the floodwave peak, and have as their X-co-ordinate the ratio of downstream distance (from the dam to a selected cross-section) to a distance parameter (X_c). The Y-co-ordinate of the curves used in predicting peak downstream flows is the ratio of the peak flow at the selected cross-section to the computed peak flow at the dam. The distinguishing characteristic of each member of a family is the ratio (V^*) of the volume in the reservoir to the average flow volume in the downstream channel. Thus it may be seen that to predict the peak flow of the floodwave at a downstream point, the desired distinguishing characteristic of the curve family and member must be determined. This determination is based on the calculation of the Froude number (F_c) and the volume ratio parameter (V^*). To specify the distance in dimensionless form, the distance parameter (X_c) in feet is computed as follows:

$$X_c = 6 \text{ VOL}/[\bar{A}(1 + 4(0.5)^{m + 1})] \tag{22}$$

in which VOL is the reservoir volume (acre-ft), m is a shape factor for the prismatic routing reach, and A is the average cross-section area in the routing reach at a depth corresponding to the height of the dam. The volume parameter (V^*) is simply $V^* = \text{VOL}/(\bar{A}_c X_c)$ in which $\bar{A}_c$ represents the average cross-sectional area in the routing reach at the average maximum depth produced by the routed flow. The Froude Number (F_c) is simply $F_c = V_c/(g D_c)^{0.5}$ where V_c and D_c are the average velocity and hydraulic depth, respectively, within the routing reach. Further details on the computation of the dimensionless parameters can be found elsewhere (Wetmore and Fread, 1984). Using families of curves similar to Figure 3, the routed peak discharge can be obtained. The corresponding peak depth is computed from the Manning equation using an iterative method since the wetted area and hydraulic radius are nonlinear functions of the unknown depth.

The time of occurrence of the peak flow at a selected cross-section is determined by adding the time of failure to the peak travel time from the dam at that cross-section. The travel time is computed using the kinematic wave velocity which is a known function of the average flow velocity throughout the routing reach. The times of first flooding and 'de-flooding' of a particular elevation at the cross-section may also be determined within SMPDBK. Further description of the computational procedure for determining these times, as well as the time of peak flow, may be found elsewhere (Wetmore and Fread, 1984).

Testing and verification

The SMPDBK model was compared with the DAMBRK model in several theoretical applications where backwater effects were negligible. The average difference between the two models was 10%-20% for predicted flows and travel times with depth differences of less than about 1 ft (0.3 m). Since the DAMBRK model is considered more accurate, the differences can be considered errors due to the simplifications of SMPDBK. The application of SMPDBK to the Teton dam breach is shown in Figure 1, and its application to the Buffalo Creek 'coal waste' dam breach is shown in Figure 2. In each case, the peak discharge profile computed with DAMBRK, and the observed peak flows are shown for comparison. On-going research and development concerning the SMPDBK model has resulted in the following improvements: (a) the interactive data input has been considerably improved allowing real-time editing capabilities and creation of a permanent data file; (b) an option to use an existing 'batch' input file; (c) peak depths are computed using the original cross-section properties as width-elevation tables rather than using a fitted 'power function' curve to replace the table (this eliminates possible significant fitting errors); (d) extension of the routing curves for Froude numbers above 0.75 by using a nonlinear extrapolation procedure. Current research is directed at possibly including the effects of downstream bridge constrictions.

BREACH

This model (Fread, 1984b) predicts the outflow hydrograph from a breached dam and the breach size, shape, and time of formation of a breach in earthen/rockfill dams where the breach may be initiated by either piping or overtopping. The dam can be man-made with either homogeneous fill or fill with a distinctive central core. The downstream face may be grass covered or bare. The model utilizes the principles of soil mechanics, hydraulics, and sediment transport to simulate the erosion and bank collapse processes which form the breach. Reservoir inflow, storage, and spillway characteristics, along with the geometrical and material properties of the dam (D_{50} size, cohesion, internal friction angle, porosity, and unit weight) are utilized to predict the outflow hydrograph. The essential model components are described as follows.

Reservoir level computation

Conservation of mass is used to compute the reservoir water surface elevation (H) due to the influence of a specified reservoir inflow hydrograph (Q_i), spillway overflow (Q_{sp}) as determined from a spillway rating table, broad-crested weir flow (Q_0) over the crest of the dam, broad-crested weir flow (Qb) through the breach, and the reservoir storage characteristics described by a surface area (S_a)-elevation table. Letting ΔH represent the change in reservoir level during a small time interval (Δt), the conservation of mass requires the following relationship:

$$\Delta H = \frac{0.0826 \, \Delta t}{S_a} (\bar{Q}_i - \bar{Q}_b - \bar{Q}_{sp} - \bar{Q}_o) \qquad (23)$$

in which the bar ($\bar{}$) denotes the average value during the Δt time interval. Thus, the reservoir elevation (H) at time (t) can easily be obtained since $H = H' + \Delta H$, in which H' is the reservoir elevation at time ($t - \Delta t$). If the breach is formed by piping, a short-tube, orifice flow equation is used instead of a broad-crested weir flow equation, i.e.

$$Q_b = 3 A_b (H - h_b)^{0.5} \qquad (24)$$
$$\text{(broad-crested weir flow)}$$

$$Q_b = A_b [2g(H - h_p)/(1 + fL1/D]^{0.5} \qquad (25)$$
$$\text{(orifice flow)}$$

in which A_b is the area of flow over the weir or orifice area, h_b is the elevation of the bottom of the breach at the upstream face of the dam, h_p is the specified centre-line elevation of the pipe, f is the Darcy friction factor which is dependent on the D_{50} grain size, L is the length of the pipe, and D is the diameter or width of the pipe.

Breach width

Initially the breach is considered rectangular with the width (B_0) based on the assumption of optimal channel hydraulic efficiency, $B_0 = B_r Y$, in which Y is the critical depth of flow at the entrance to the breach; i.e. $Y = 2/3(H - h_b)$. The factor B_r is set to 2 for overtopping and 1 for piping. The initial rectangular-shaped breach can change to a trapezoidal shape when the sides of the breach collapse due to the breach depth exceeding the limits of a free-standing cut in soil of specified properties of cohesion (C), internal friction angle (ϕ), unit weight (γ), and existing angle (θ') that the breach cut makes with the horizontal. The collapse occurs when the effective breach depth (d') exceeds the critical depth (d_c), i.e.

$$d_c = 4C \cos \phi \, \sin \theta'/[\gamma - \gamma \cos(\theta' - \phi)] \qquad (26)$$

The effective breach depth (d') is determined by reducing the actual breach depth (d) by $Y/3$ to account for the supporting influence of the water flowing through the breach. The θ' angle reduces to a new angle upon collapse which is simply $\theta = (\theta' + \phi)/2$. The model allows up to three collapses to occur.

Breach erosion

Erosion is assumed to occur equally along the bottom and sides of the breach except when the sides of the breach collapse. Then, the breach bottom is assumed not to continue to erode downward until the volume of collapsed material along the length of the breach is removed at the rate of sediment transport occurring along the breach at the instant before collapse. After this characteristically short pause, the breach bottom and sides continue to erode. Material above the wetted portion of the eroding breach sides is assumed to simultaneously collapse as the sides erode. Once the breach has eroded to the specified bottom of the dam, erosion continues to occur only along the sides of the breach. The rate at which the breach is eroded depends on the capacity of the flowing water to transport the eroded material. The Meyer-Peter and Muller sediment transport relation as modified by Smart (1984) for steep channels is used, i.e.

$$Q_s = 3.64(D_{90}/D_{30})^{0.2} \frac{D^{2/3}}{n} P S^{1.1} (DS - 0.0054 D_{50} \tau_c) \qquad (27)$$

in which Q_s is the sediment transport rate, D_{90}, D_{30} and D_{50} are the grain sizes in (mm) at which 90%, 30%, and 50% respectively of the total weight is finer, D is the hydraulic depth of flow computed from Manning's equation for flow along the breach at any instant of time, S is the breach bottom slope which is assumed to always be parallel to the downstream face of the dam, and τ is Shield's critical sheer stress that must be exceeded before erosion occurs. The incremental increase in the breach bottom and sides (ΔH_c) which occurs over a very short interval of time is given by:

$$\Delta H_c = Q_s \Delta t / [P\ L(1 - p)] \tag{28}$$

in which P is the total perimeter of the breach, L is the length of the breach through the dam, and p is the porosity of the breach material.

Computational algorithm

The sequence of computations in the model are iterative since the flow into the breach is dependent on the bottom elevation of the breach and its width while the breach dimensions are dependent on the sediment transport capacity of the breach flow; and the sediment transport capacity is dependent on the breach size and flow. A simple iterative algorithm is used to account for the mutual dependence of the flow, erosion, and breach properties. An estimated incremental erosion depth (ΔH_c) is used at each time step to start the iterative solution. This estimated value can be extrapolated from previously computed values. Convergence is assumed when ΔH_c computed from Equation (28) differs from ΔH_c by an acceptable specified tolerance. Typical applications of the breach model require less than two minutes on microcomputers with a fast arithmetic processor. The computations show very little sensitivity to a reasonable variation in the specified time step size. The model has displayed a lack of numerical instability or convergence problems.

Teton application

BREACH was applied to the piping initiated failure of the Teton earthfill dam which breached in June 1976, releasing an estimated peak discharge of 2.3 million cfs (65,128 cm) having a range of 1.6 to 2.6 million cfs. The simulated breach hydrograph is shown in Figure 4. The computed final top breach width of 650 ft (214.7 m). The computed slide slope of the breach was 1:1.06 compared to 1:1.00. Additional information on this and another application of BREACH to the naturally formed landslide dam on the Mantaro river in Peru, which breached in June 1974, can be found elsewhere (Fread, 1984b). The model has also been satisfactorily verified with the piping-initiated failure of the 28 ft (8.5 m) high Lawn Lake dam in Colorado (Jarrett and Costa, 1982). BREACH will continue to be tested as data becomes available.

Summary

Three NWS models for predicting the flooding due to dam failures are summarized. The BREACH model can aid the hydrologist/engineer in determining the properties of the piping or overtopping initiated breach of an earthen dam. This information can be used in conjunction with historical

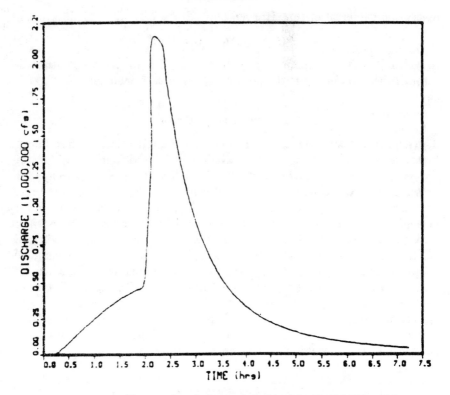

FIGURE 4 *Teton outflow hydrograph produced by BREACH model*

breach data to create the dam breach hydrograph and route it through the downstream channel/valley using the complex DAMBRK model or the simplified SMPDBK model. The choice of either the DAMBRK or SMPDBK model is influenced by the available time, data, computer facilities, modelling experience, and required accuracy for each dam break analysis. Complexities in the downstream channel valley such as highway/railway embankment-bridges, significant channel contrictions, levée overtopping, flow volume losses, downstream dams, weirs, and lakes require the DAMBRK model to be used rather than the SMPDBK model since the latter model ignores such factors.

REFERENCES

DAVIES, W.E., BAILY, J.F. and KELLY, D.B. (1975) West Virginia's Buffalo Creek flood: a study of the hydrology and engineering geology. *Geological Survey,* Circular 667, US Geological Survey, 1975, 32 pp

DELONG, L.L. (1985) Extension of the unsteady one-dimensional open channel flow equations for flow simulation in meandering channels with flood plains. *Selected Papers HydroSci*, 101–105

FREAD, D.L. and HARBAUGH, T.E. (1973) Transient hydraulic simulation of breached earth dams. *J Hydraul Div*, (Am Soc Civil Eng) **99**: 139–154

FREAD, D.L. (1977) The development and testing of a dam-break flood forecasting model. *Proceedings, dam-break flood modeling workshop*. US Water Resourses Council, Oct. 18–20, 1977. Washington DC, 32 pp

FREAD, D.L. (1983) Computational extensions to implicit routing models. *Proceedings of conference on frontiers in hydraulic engineering*. ASCE, Massachusetts Institue of Technology, Cambridge, Massachusetts, Aug. 1983, 343–348

FREAD, D.L. (1984a) DAMBRK: the NWS dam-break flood forecasting model. Hydrologic Research Laboratory, National Weather Service, Silver Spring, Maryland, 56 pp

FREAD, D.L. (1984b) A breach erosion model for earthen dams. *Proceedings of specialty conference on delineation of landslides, flash flood, and debris flow hazards in Utah*. Utah State University, June 15, 1984, 30 pp

FREAD, D.L. (1985) Channel routing. In: *Hydrological Forecasting*. Anderson, M.G. and Burt, T.P. (Eds). ch 14, John Wiley and Sons, Chichester, 437–503

FREAD, D.L. (1988) The NWS DAMBRK model: theoretical background/user documentation. Hydrologic Research Laboratory, National Weather Service, Silver Spring, Maryland, 320 pp

FROELICH, D.L. (1987) Embankment – dam breach parameters. *Proceedings of 1987 conference on hydraulic engineering*. ASCE, New York, Aug. 1987, 570–575

JARRETT, R.D. and COSTA, J.E. (1982) Hydrology, geomorphology, and dam-break modelling of the July 15, 1982, Lawn Lake Dam and Cascade Lake Dam failures, Larimer County, Colorado. US Geological Survey, Open File Report 84–62, 109 pp

JOHNSON, F.A. and ILLES, P. (1976) A classification of dam failures. *Water Power Dam Constr*, 43–45

RAY, H.A., KJELSTRON, L.C., CROSTHWAITE, E.G. and LOW, W.H. (1976) The flood in southeastern Idaho from Teton dam failure of June 5, 1976. *Open File Report*, US Geological Survey, Boise, Idaho

SAMUELS, P.G. (1985) Models of open channel flow using Preissmann's scheme. Presented at Cambridge University, England, Sept. 24–26, 91–102

SMART, G.M. (1984) Sediment transport formula for steep channels. *J Hydraul Div* (Am Soc Civil Eng) **110**: 267–276

WETMORE, J.N. and FREAD, D.L. (1984) The NWS simplified dam break flood forecasting model. Federal Emergency Management Agency (FEMA), 122 pp

Study of Earth Dam Erosion due to Overtopping

F. MACCHIONE[1] and B. SIRANGELO[2]

[1]Department of Soil Defence, University of Calabria,
[2]Institute of Civil Engineering, University of Salerno, Italy

Introduction

THE PROBLEM of dam failures has always been of great importance because of their disastrous effects. Therefore, it has always given rise to a particular interest among hydraulic engineers in order to estimate downstream valley areas exposed to the hazard of inundation.

Of great interest is the study of earthen embankment failures which are by far the most common type of dam in the world. A recent report has shown that the frequency of failure of such dams is about four times greater than that observed for concrete and masonry dams (Lebreton, 1985).

In Italy two recent events have drawn engineers' attention to the necessity of increasing the present knowledge about this matter:

- the failure of two settling reservoirs of a mining plant in Val di Stava, which caused heavy loss of human life; and
- the blocking of an upper reach of the Adda river in Valtellina, due to a landslide. In the latter event, overtopping of the embankment was kept under control by a Technical Commission of the Ministry of Civil Protection.

The present paper deals with an analysis of the interaction between reservoir routing and earth embankments in order to simulate the erosion process and evaluate and outflow hydrograph.

Failure of Earthen Dams

Among the main causes of dam failures, besides floods exceeding the spillway capacity, may be included settlements, foundation failures, cracks, embankment slips and landslides falling into the reservoir. Many such conditions are due to earthquakes and they can also be joined to water reservoir waves (Sherard et al., 1963; Johnson and Illes, 1976; Seed et al., 1980).

Apart from their causes, failures can be followed by a water release either on the top of the dam (overtopping) or through the embankment (piping). In the latter case, seeping water makes a free path through the dam; this increases in size until the material above this hole collapses and the overtopping begins.

Observations of past dam failures have indicated that the breach shape is usually triangular and its width and depth grow during overtopping. In the case of earthfill dams, the breach can generally grow until it reaches the natural ground, which is less erodable, and then it develops laterally so its final shape will be trapezoidal. More complex and erratic is the breach development in earth dams with protective concrete surface layers and core walls (McDonald and Langridge-Monopolis, 1984).

It is important to point out that observed times for earth dam erosion show that breach growth is not a fast process. It is just this gradual character that distinguishes the failures of earthen embankments from those of concrete dams.

Model Formulation

In the last 10 years several breach flood wave models have been developed with the purpose of simulating the outflow hydrograph and routing this hydrograph through the downstream valley. Concerning the breach simulation it is possible to distinguish the following methods (Wurbs, 1987):

– instantaneous complete removal of the dam;
– instantaneous partial breach of the dam;
– breach whose growth is fixed with time;
– breach whose growth is predicted using an erosion model.

The first two methods may be appropriate respectively for a concrete arch dam and for a concrete gravity dam, but they seem to be too conservative and unrealistic for earth embankments. The assumption that breach dimension grows with time, usually according to a linear law, appears more likely. However, the fourth method, adopted in the present paper, provides a more realistic representation of the erosion process.

Referring to the definition sketch in Figure 1, the flow of water over the dam can be described by the one-dimensional De Saint Venant equations:

$$\frac{\partial Q}{\partial x} + \frac{\partial A}{\partial t} + \frac{\partial A_d}{\partial t} = 0 \qquad (1)$$

$$\frac{\partial Q}{\partial t} + \frac{\partial}{\partial x}\left(\frac{Q^2}{A}\right) + gA\left(\frac{\partial Z}{\partial x} + S_f\right) = 0 \qquad (2)$$

where Q is the discharge, A is the cross-sectional area of water, A_d is the cross-

213

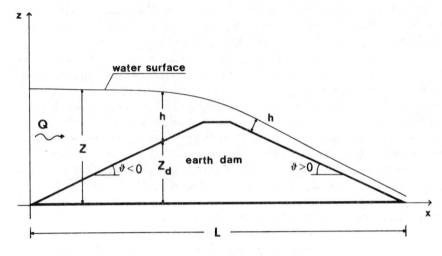

FIGURE 1 *Definition sketch*

sectional area of bed deposit, Z is the water surface elevation, S_f is the friction slope, g is the acceleration due to gravity and x and t are, respectively, the longitudinal space co-ordinate and the time. The friction slope can be expressed with one of empirical equations for open channel resistance: $S_f = Q^2/K^2$, where K is the conveyance factor.

Together with the continuity and dynamic equations for water flow a continuity equation is used for sediment:

$$\frac{\partial G_s}{\partial x} + \gamma_s(1-\lambda)\frac{\partial A_d}{\partial t} + \frac{\partial}{\partial t}(A\,C_s) = 0 \qquad (3)$$

where G_s is the solid discharge, C_s is the suspended sediment concentration and γ_s and λ are, respectively, the specific weight and the porosity of sediment on the bed.

The set of partial differential Equations (1), (2), (3) links the three unknown functions Q, Z and A_d with independent variables x and t and can be solved with suitable boundary and initial conditions. At the upstream section it is possible to give three kinds of boundary conditions: the discharge $Q(t)$, the water surface elevation $Z(t)$ or a stage-discharge relationship.

In dynamic reservoir routing such an upstream condition is usually given by a continuity equation applied to reservoir storage volume W:

$$\frac{dW}{dt} = Q_i - Q_s - Q \qquad (4)$$

where Q_i is the inflow to the reservoir, Q_s is the outflow due to spillways and W is linked to Z by a storage-elevation relationship.

Since the flow will be subcritical upstream and supercritical downstream a further condition is needed:

$$Fr^2 = \frac{Q^2 B}{gA^3} = 1 \qquad (5)$$

where B is the width of cross-section at free surface elevation and Fr is the Froude number. Equation (5) will be the second boundary condition for the upstream subcritical flow. The computed values of Q and Z at the last subcritical flow section will give the two boundary conditions for the downstream supercritical flow.

As regards erosion, the boundary condition in the first upstream section can be simply given by $A_d = 0$.

Water profile at $t = 0$ is evaluated as a steady, gradually varied flow, although this assumption is only a rough approximation of the actual initial condition.

Sediment Transport Equation

The evaluation of the solid discharge G_s can be obtained from one of the several sediment transport formulas developed for open channels (Vanoni, 1975).

However, it is important to point out that the conditions in which movement of sediment occurs during erosion of dams are very different from those for which the formulae have been calibrated.

In fact, during overtopping water flow is far from uniform, sediment transport is not in equilibrium conditions and shear stress can reach extremely high values.

Furthermore, for embankments built up with a clay core, during overtopping a mixture of cohesive and cohesionless material is eroded. In such conditions the sediment transport process needs further study.

Numerical Approach

The simultaneous integration of the three partial differential Equations (1), (2), (3) is rather difficult.

However, in the case here analyzed it is possible to uncouple the Equation (3) from the system. Therefore, the solution of the system can be found solving, for each computational time step, first the De Saint Venant equations and then the sediment continuity equation (Chen et al., 1975).

Concerning downstream supercritical flow, moreover, the unsteadiness of the water movement is due mainly to erosion so, during each time step, it can

215

be considered as steady and computed, from critical section to downstream, as follows (Chow, 1959):

$$\frac{dh}{dx} = \frac{\tan\theta - S_f/\cos\theta}{\cos\theta - h\sin\theta\, d\theta/dh - Fr^2} \tag{6}$$

where h is the water depth and $\theta = \theta(x)$ is the bottom slope (Figure 1).

During the erosion process the condition (5) has always been located in the section having the maximum bottom elevation. Such a condition has allowed to reproduce the actual behaviour of water flow.

Numerical integration of the Equations (1), (2) and (3) has been made using the finite difference method. The De Saint Venant equations and the sediment continuity equation have been discretized and linearized, respectively, according to the Preissmann implicit scheme and the linear centre implicit method.

Particular attention has been paid to the choice of space and time steps, as well as to the choice of the weighting coefficient of the Preissmann scheme, in order to avoid numerical instability in the reach where the Froude number is close to one. Moreover, in the downstream reach the best results have been achieved by introducing a smoothing technique for bottom profile.

Model Verification

The effectiveness of the mathematical model suggested here has been checked on the basis of laboratory experiments carried out in the EDF National Hydraulic Laboratory, Chatou, France. The experience here considered concerns the erosion of a sand-dyke model described by Benoist and Nicollet (1983).

It has been found that the ability of the mathematical model to simulate the experimental data strongly depends on sediment transport formula adopted for G_s estimation. The best results have been achieved by neglecting suspended transport and adopting the Meyer-Peter and Mueller (1948) formula:

$$\phi = 8(\tau\star - \tau\star_c)^{3/2} \tag{7}$$

where $\tau\star$ and $\tau\star_c$ are, respectively, the dimensionless shear stress and its critical value, and where:

$$\phi = \frac{G_s/B}{\gamma_s d\sqrt{\gamma' g d}} \tag{8}$$

with d mean sediment diameter, $\gamma' = (\gamma_s - \gamma)/\gamma$ and γ water specific weight. Since the slope of sand-dyke profile is not small, the critical dimensionless shear stress has been modified according to:

$$\tau\star_c = \tau\star_{co}\cos\theta(1 - \tan\theta/\tan\varphi) \tag{9}$$

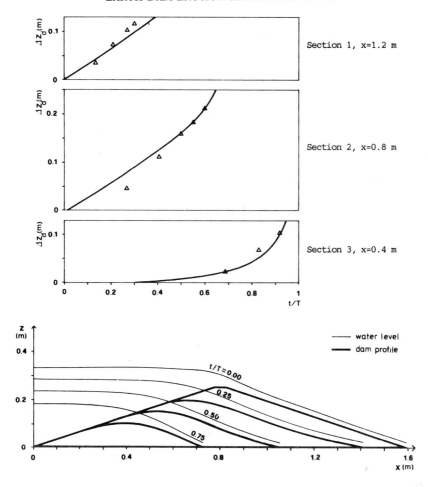

FIGURE 2 (a) – Comparison between model results and experimental data, T = complete erosion time, ΔZ_d = sand-dyke erosion. $\triangle$, observed (Benoist and Nicollet, 1983), —, simulated (with Meyer-Peter and Mueller formula). Top: $x = 1.2$ m; middle: $x = 0.8$ m; bottom: $x = 0.4$ m. (b) – Simulated development of water levels (fine line) and dam profiles (heavy line)

where φ is the submerged angle of repose of the material and $\tau*_{co}$ is the critical dimensionless shear stress on horizontal bed.

Figure 2(a) compares, on three dam sections, model outcome and experimental data. A good agreement between computed and observed erosion is obtained except for the middle dam section at the beginning of erosion. This is, perhaps, due to initial conditions employed in the mathematical model that might not reproduce the experimental conditions exactly.

217

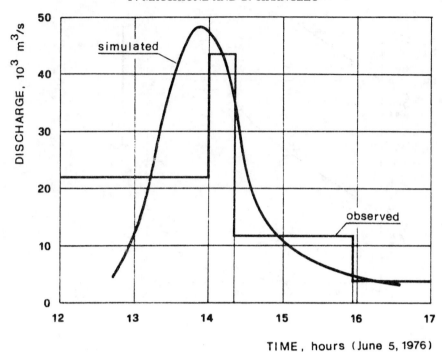

FIGURE 3 *Simulation of the outflow hydrograph due to Teton Dam Failure*

The simulated development of water and dam profiles are depicted in Figure 2(b).

Model Application

The model has been applied to simulate the outflow hydrograph of the failure of Teton Dam, Idaho, USA. Field data, including observations of reservoir elevation during the failure, have been reported by Ray and Kjelstrom (1978).

Figure 3 shows the computed outflow hydrograph compared with the hystograph obtained with reservoir volume differentials and corresponding mean discharge. The agreement is very satisfactory and the simulated peak discharge according to the value estimated by Balloffet and Scheffler (1982).

Model results have been achieved by multiplying the solid discharge given by Meyer-Peter and Mueller formula by a factor of 2.4. The value of this factor, however, is strongly affected by the assumption, somewhat arbitrary, relative to breach geometry and then must be considered as a calibration parameter without a clear physical meaning.

Conclusions

The phenomenon of earth dam failure due to overtopping is described, analyzing the unsteady flow of water over the embankment and the resulting erosion process.

The proposed mathematical model has been verified on the basis of laboratory experiments reported in the literature and it has given a good fitting of experimental data.

The application of the model to an actual dam failure has provided a satisfactory simulation of the observed outflow hydrograph.

Therefore, the model can be considered as a useful tool for earth dam breaching analysis and, even if depending on a careful calibration, can provide a realistic outflow hydrograph prediction.

REFERENCES

BALLOFFET, A. and SCHEFFLER, M.L. (1982) Numerical analysis of the Teton Dam failure flood. *J Hydraul Res* **20**: 317–328

BENOIST, G. and NICOLLET, G. (1983) Rupture progressive des barrages en terre. Proceedings XX IAHR Congress, Moscow, Vol II, 464–470

CHEN, Y.H., HOLLY, F.M., MAHMOOD, K. and SIMONS, D.B. (1975) Transport of material by unsteady flow. In: *Unsteady flow in open channels.* Yevjevich, V. and Mahmood, K. (Eds) Vol I. Water Resources Publications, Fort Collins, Colorado, pp. 313–365

CHOW, V.T. (1959) *Open channel hydraulics.* McGraw-Hill, New York, 680 pp

JOHNSON, F.A. and ILLES, P. (1976) A classification of dam failures. *Water Power Dam Constr* **28**: 43–45

LEBRETON, A. (1985) Les ruptures et accidents graves de barrages de 1964 à 1983. *La Houille Blanche* **6/7**: 529–544

McDONALD, T.C. and LANGRIDGE-MONOPOLIS, J. (1984) Breaching characteristics of dam failures. *Proc ASCE J Hydraul Eng* **110**: 567–586

MEYER-PETER, E. and MUELLER, R. (1948) Formulas for bed-load transport. Proceedings II IAHR Meeting, Stockholm, pp. 39–64

RAY, H.A. and KJELSTROM, L.C. (1978) The flood in southeastern Idaho from the Teton Dam failure of June 5, 1976. U.S. Geological Survey, Open File Report 77–765.

SEED, H.B., MAKDISI, F.I. and DE ALBA, P. (1980) The performance of earth dams during earthquakes. *Water Power Dam Constr* **32**: 17–27

SHERARD, J.L., WOODWARD, R.T., GIZIENSKI, S.F. and CLEVENGER, W.A. (1963) Earth and earth-rock dams. Wiley, New York, 725 pp

VANONI, V.A. (ED.) (1975) Sedimentation engineering. *ASCE Man Rep Eng Prac* **54**: 745

WURBS, R.A. (1987) Dam-breach flood wave models. *Proc ASCE, J Hydraul Eng* **113**: 29–46

Hazardous Hydrologic Consequences of Volcanic Eruptions and Goals for Mitigative Action: An Overview

—

THOMAS C. PIERSON

Water Resources Division,
US Geological Survey, Cascades Volcano Observatory,
Vancouver, WA 98661, USA

Introduction

VOLCANOES occur in many shapes and sizes and in a variety of geologic settings. They also vary greatly in the frequency and violence of their eruptions. Basaltic shield volcanoes, such as those in Hawaii, generally have effusive eruptions, with relatively little explosive activity. Although some may achieve topographic relief in excess of 4,000 m; shield volcanoes have relatively flat slopes and are composed predominantly of moderately strong rock layers resulting from congealed lava flows. Hazardous hydrologic processes occur relatively infrequently at such volcanoes.

Stratovolcanoes, however, commonly achieve topographic relief in excess of 4,000 m, have steep and erodible slopes, are partly or predominantly constructed of weakly consolidated or loose rock debris, and have the potential for violently explosive eruptions. These tall, steep, volcanic cones can be further weakened by hydrothermal alteration to the point of mechanical failure. In addition, they are commonly mantled at their summits by thick accumulations of ice and snow, even at tropical latitudes, and many have crater lakes. Such volcanoes are commonly susceptible to a number of hazardous hydrologic processes.

Catastrophic releases or production of large volumes of water, ice and snow, and/or rock debris (commonly water-saturated) are the most hazardous potential hydrologic consequences of volcanic eruptions. But it must be emphasized that some of these consequences occur not only during or shortly after eruptions; they can occur decades, centuries, or even millenia after the last volcanic eruption. Once released, large masses of flowing rock debris and water can erode and incorporate more sediment and water, growing steadily in volume with distance downslope. These masses of sediment and water can flow tens to hundreds of kilometres as debris avalanches, debris flows, or floods, depending on the ratio of water to sediment. Events occurring during non-eruptive periods are particularly dangerous because they can occur without warning and may be impossible to predict.

Catastrophic Release Mechanisms

The mechanisms that trigger large sediment-water flows during eruptions or during volcanically quiescent periods are hydrodynamic in scope. The basic categories of mechanisms, which may operate singly or in combination, include (1) release of water already being stored on or near the volcano, (2) production of water by melting snow and ice, (3) slope failure of part of the volcanic cone or its snow/ice cover owing at least partly to seepage forces from meltwater, perched groundwater, or hydrothermal water, or (4) production of fragmental rock debris during an eruption that will mix with water as it travels downslope. Specific triggering mechanisms are listed in Table 1. These have been discussed and contrasted in review papers by Neall (1976) and Major and Newhall (1989).

Types and Characteristics of Sediment-Water Flows

Once a large mass of water, snow and ice, or rock debris is released on a slope, the mass will immediately begin to flow. When it encounters erodible material on a sufficiently steep slope, whether that material is snow, soil, or water in a channel, erosion and mixing will occur, and the flow will increase in volume by incorporation of that material. Volume increases of over an order of magnitude can occur (Gallino and Pierson, 1985).

The rheological behaviour of the flowing mass depends on the sediment/water ratio. A broad spectrum of flow types is possible; a few documented examples are given in Table 2. Relatively dry (nonliquefied) granular flows of rock debris have been classified as debris avalanches by Varnes (1978) and Pierson and Costa (1987). Particulate flows of snow and ice also are termed avalanches, and snow avalanches triggered by volcanic eruptions can contain substantial amounts of rock debris as well (Waitt et al., 1983; Janda et al., 1986). Flows involving viscous, liquefied slurries of rock debris and water are termed debris flows (Varnes, 1978; Pierson and Costa, 1987). Such flows have significant shear strength and are capable of suspending large boulders in the fluid. Large flows involving mostly water have long been termed floods.

TABLE 1 *Mechanisms capable of releasing or producing large masses of water, snow and ice, or rock debris on or near volcanoes*

Mechanism	Hazardous hydrologic consequences	Examples	References
Volcanic mechanisms[1]			
Endogeneous			
Earthquake (volcanic)	Landslide; snow/ice avalanche; liquefaction of avalanche deposit	Nevado del Ruiz, 1845 Mount St Helens, 1980	Mojica et al., 1985; Herd et al., 1986. Fairchild, 1985; Voight et al., 1981, 1983.
Forces on volcanic cone from magma injection prior to eruption or during eruption	Landslide	Mount Rainier, 5700yBP Mount St Helens, 1980	Crandell, 1971. Voight et al., 1981, 1983.
Eruption of mud and rock debris during phreatic eruptions	Debris flow	Usu, 1910 Dieng, 1928, 1979	Omori, 1911; Oinouye, 1917. Koninklijk Magnetisch en Meteorologisch Obs de Batavia, 1929; Sudrajat and Siswowidjoyo, 1981.
Exogeneous			
Pyroclastic surge/blast	Snow/ice avalanche; rapid surficial melting of ice and snow	Mount St Helens, 1980 Mount St Helens, 1982 Nevado del Ruiz, 1985	Janda et al., 1981; Pierson, 1985; Major and Voight, 1986; Scott, 1988. Wiatt et al., 1983; Pierson and Scott, 1985. — Janda et al., 1986; Lowe et al., 1986; Naranjo et al, 1986;
Pyroclastic flow		Nevado del Ruiz, 1985	— Thouret et al., 1987.
Lava flow	Rapid melting of ice and snow[2]	Llaima, 1979 Villarrica, 1984	Moreno-Roa et al., 1981. Gonzales-Ferran, 1985.
Subglacial eruption	Rapid melting of ice, glacier outburst flood	Katla, 1918 Deception Island, 1967–69	Thorarinsson, 1975; Jonsson, 1982. Baker 1969; Baker et al., 1969.
Eruption through crater lake	Water and debris thrown onto volcano's flank	Ruapehu, 1969 Kelut, 1919, 1966	Healy et al., 1978. Suryo and Clarke, 1985.
Volcanic debris damming streams	Dam failure by overtopping, dam-break flood	Asama, 1783 El Chichon, 1982	Aramaki, 1956. Silva et al., 1982.
Eruption-induced rainfall	Intense rainfall localized on volcano's flanks	Mayon, 1968	Moore and Melson, 1969
Nonvolcanic mechanisms			
Endogeneous			
Earthquake (tectonic)	Landslide; snow/ice avalanche	Iriga, 1628 (probable) Mt Ontake, 1984	Agulia et al., 1986. Inokuchi, 1985; Ashida and Egashira, 1986.

TABLE 1 (Continued)

Mechanism	Hazardous hydrologic consequences	Examples	References
Hydrothermal alteration of rock, weathering	Landslide	Mt Rainier, 600 yBP	Crandell, 1971.
Exogenous			
Englacial or subglacial storage of meltwater from fumarolic or climatic melting	Glacial outburst floods	Grimsvotn, 1972, 1976, 1982	Thorarinsson, 1953; Bjornsson, 1974, 1975; Bjornsson and Kristmannsdottir, 1984.
		Mt Rainier, 1947	Richardson, 1968.
Formation of unstable natural dams: In craters, by landslides, by glacial moraines	Dam-break floods	Mount St Helens, 1980	Dunne and Leopold, 1981. Jennings *et al.*, 1981.
		Ruapehu, 1953	O'Shea, 1954.
		Broken Top, 1966	Costa and Schuster, 1988.
		Kelut, 1875	Suryo and Clarke, 1985.
		Mt Hood, 1980	Gallino and Pierson, 1985.
Heavy rainfall	Floods, debris flows	Mayon, 1984, 1985	Okkerman *et al.*, 1985; Umbal, 1989; Rodolfo *et al.*, 1989.
		Merapi, 1930	Suryo and Clarke, 1985.
		Galunggung, 1982	Suryo and Clarke, 1985; United Nations, 1985.
		Irazu, 1963–65	Waldron, 1967.
		Sakurajima, 1955–present	Watanabe and Ikeya, 1981.
		Mt Hood, 1980	Gallino and Pierson, 1985

[1] Mechanisms occurring during or within a few days after the eruption.
[2] Lava flows generally do not melt snow rapidly enough to trigger consequential debris flows (Major and Newhall, 1989), but lava flowing on steep slopes may brecciate, resulting in more efficient mixing and heat transfer, and consequently more rapid melting.

Hyperconcentrated floods (an intermediate category between normal floods and debris flows) entrain sufficient sediment to provide a small amount of shear strength and to dampen turbulence (Pierson and Costa, 1987). Hyperconcentrated floods commonly occur as the more dilute runout flows of volcanic debris flows (Janda *et al.*, 1981; Pierson and Scott, 1985; Scott, 1988).

Volcanic debris avalanches begin as landslides; they can mobilize unimaginable volumes of debris and travel great distances at extremely high velocities (Siebert, 1984; Siebert *et al.*, 1987; Table 2). Large debris avalanches were not widely recognized as a hazard from stratovolcanoes until 1980, when one comprising 2.5 to 2.8 km^3 of debris rushed 25 km down a valley in 10 minutes at Mt. St. Helens (Voight *et al.*, 1983). An exceptionally large debris avalanche of Pleistocene age occurred at Mt. Shasta, California; it mobilized 45 km^3 of debris, covered an area of about 675 km^2, and extended 49 km away from the present summit of the volcano (Crandell, 1988). Debris avalanches from volcanoes near coasts or on islands can extend their destruction even farther by generating tsunamis (large gravitational waves). One such wave

223

TABLE 2 *Examples of origins and characteristics of different types of large sediment-water flows originating from volcanoes*

Flow type	Location, date	Origin	Volume (m³)	Velocity (m/s)	Peak discharge (m³/s)	Distance travelled (km)	Reference
Debris avalanche (evolved into debris flow)	Mt Ontake, 1984	Earthquake-triggered landslide	3.6×10^7	36 (max) 23 (mean)	—	13.5	Inokuchi, 1985; Ashida and Egashira, 1986.
Debris avalanche	Mt Shasta	—	4.5×10^{10}	—	—	49	Crandell, 1988.
Debris avalanche	Mount St Helens, 1980	Landslide	$2.5\text{–}2.8 \times 10^9$	70 (max) 42 (mean)	—	25	Voight et al., 1983.
Debris flow	Mount St Helens, 1980 (North Fork Toutle River)	Remobilization of static debris-avalanche deposit	1.4×10^8 1.3×10^8	5.3 (front) 10 (peak) 7 (peak) 1.7 (front)	7,200 6,600 6,000	at 4.5 at 26 at 40 (125 total)	Janda et al., 1981; Fairchild, 1985.
Debris flow	Mount St Helens, 1980 (South Fork Toutle River)	Rapid snowmelt by pyroclastic surge	1.3×10^7 8.3×10^6	33 (peak) 4–8 (peak) 2.4 (front)	6,800 3,800	at 4 at 44 (59 total)	Cummans, 1981; Fairchild, 1985.
Debris flow	Nevado del Ruiz (Rio Azufrado)	Rapid snowmelt by pyroclastic flow and surge; snow avalanches	1.2×10^7 5.5×10^7	14.6 (peak) 12.0 (peak) 6 (front)	48,000 27,100	at 10 at 69 (90 total)	Pierson et al., 1986.
Flood	Broken Top, 1966	Moraine-dam failure	1.9×10^5	—	71	9	Costa and Schuster, 1988.
Flood	Grimsvotn, 1954	Glacier outburst flood	$3\text{–}3.5 \times 10^9$	—	10,000	—	Bjornsson, 1974.

generated by a debris avalanche from Unzen volcano in 1792 killed at least 4,300 people in Japan (Schuster and Crandell, 1984).

Debris flows at volcanoes can be caused by landslides, floods, wet snow and ice avalanches, intense rainfall, hot pyroclastic debris ejected onto snow, or (in special situations) by direct ejection of mud slurries during eruptions (Neall, 1976; Major and Newhall, 1989). Examples of these different origins are given in Tables 1 and 2. Although not as rapid as some debris avalanches, debris flows are capable of flowing much farther when confined to river valleys. Debris flows have been known to flow as fast as 40 m/s (Nakamura, 1926; Janda et al., 1981; Pierson, 1985) and as far as 125 km (Janda et al., 1981). Volcanic debris flows commonly flow between 10 and 20 m/s and are extremely hazardous to people living in confined vallleys or on alluvial fans at canyon mouths. The town of Armero, Colombia, was situated on such a fan 72 km downstream from the volcano Nevado del Ruiz. A debris flow triggered by the 13 November 1985 eruption of that volcano travelled the distance in about two hours and killed more than 23,000 people.

Floods at and in the vicinity of volcanoes can be initiated by heavy rainfall, rapid melting of snow and ice, lake outbreaks (failures of natural or man-made dams), and glacier outbursts (release of reservoirs from within or beneath glacier ice); see Tables 1 and 2. Blankets of volcanic ash on the landscape commonly result in decreased infiltration capacitites; runoff is produced more quickly and in greater volumes, and the frequency of flooding is increased. Particularly sudden and unexpected releases of water can occur from glacier outbursts and dam failures; the requisite large impoundments of water can be formed quickly during eruptions by landslides or flowage deposits. In steep erodible channels, floods may bulk up with sediment and transform into debris flows.

A persistent and far-reaching effect of explosive volcanic eruptions is the accelerated deposition of volcanic sediment in lowland valleys, which leads to channel aggradation and increased lateral migration of channels and bank erosion (Waldron, 1967; Smart, 1981; Meyer and Janda, 1986). These conditions can exacerbate normal seasonal flooding and cause extensive damage to bridges, roads, and homes built on floodplains. As was demonstrated at Mt. St. Helens following the 1980 eruption, significant aggradation can be caused both by 'instantaneous' deposition of debris-flow deposits during the eruption and by the highly accelerated deposition of fluvially eroded volcanic sediment in the decade following the eruption (Dunne and Leopold, 1981; Janda et al., 1981; Lombard et al., 1981; Vanoni, 1982; Lehre et al., 1983; Collins and Dunne, 1986). In such cases, the channel capacity required to convey rainfall runoff is partly filled by the sediment, and even small floods are forced to flow overbank. To counter channel aggradation at Mt. St. Helens, over $340 million has been spent since 1980 for channel dredging, dyking, and construction of sediment-retention dams and basins, in order to prevent downstream flooding of the lower Toutle and Cowlitz River valleys (Karl Ericksen, US Army Corps of Engineers, written communication, 1988). Significant post-eruption excess sedimentation also has been documented for Kelut volcano in Indonesia (Smart, 1981) and for Irazu volcano in Costa Rica (Waldron, 1967).

Goals for Mitigative Action

Monitoring

Volcanoes situated in populated areas that have a significant potential for catastrophic sediment-water flows can be monitored. A relatively inexpensive, 'low-tech' way to do this in densely populated regions is to pay local residents to continuously occupy safe observation posts on the volcano's flanks near dangerous channels. Warnings of flows seen (or heard at night) could be sent down to populations at risk, by radio or by acoustic means (drums made from hollowed logs are used for this purpose in Indonesia). For more dangerous or more remote volcanoes, various types of electronic or mechanical sensors can be used to detect passage of a flow or a sudden drop in the level of a crater lake (Hewson and Latter, 1976; Paterson, 1980; Childers and Carpenter, 1985). Ground-based or satellite radio telemetry can relay the warning automatically to populations downvalley, but such systems require periodic (sometimes frequent) maintenance by trained personnel.

Techniques and instrumentation routinely used by volcanologists can also be useful in hydrologic monitoring. Seismograph networks can be useful in detecting (and possibly locating) large sediment-water flows coming from volcanoes. Precise surveying, which is used to detect pre-eruption deformation of volcanoes, might also be used to detect incipient motion of large landslide blocks on the sides of volcanoes. Periodic aerial photography could show abrupt changes in drainage or seepage patterns, the changed condition of an ice cap, or even development of tension cracks or bulges, any of which could presage the occurrence of a large landslide. If resources are limited, however, it may be difficult to decide which volcano (among several threatening candidates) to monitor or how many resources to commit to a particular volcano.

Defensive strategies

When the physiographic, geologic, and hydrologic conditions of a particular volcano have been investigated, including the preserved record of the magnitude and frequency of past sediment water flows, hazard maps can be prepared (United Nations, 1977; Crandell et al., 1984). Unless a crisis situation exists, rational decisions can then be made about how to mitigate the effects of these hazards, should they occur, and hazard management plans can be developed and put into effect (Tomblin, 1982; Suryo and Clarke, 1985; United Nations, 1985).

A wide range of defensive strategies can be employed to protect people and property from the kinds of hydrologic hazards discussed in this paper. Several broad categories of defensive strategies are listed in Table 3. More extensive discussions of these and similar mitigative strategies may be found in Van Padang (1960), Gagoshidze (1969), Hewson and Latter (1976), Nasmith and Mercer (1979), Paterson (1980), Watanabe and Ikeya (1980), Smart (1981), Hungr et al. (1984), Kockelman (1984, 1986), Martin et al. (1984), Childers and Carpenter (1985), Sumaryono and Kondo (1985), Suryo and Clarke (1985), and Blong and Johnson (1986).

The choice of a defensive strategy will depend on many factors, including economic and political pressures. However, at most volcanoes in the world, the most fundamental obstacle to successful mitigation of hydrologic disasters is not specifically the lack of monitoring or the incorrect choice of a defensive strategy but rather a basic lack of understanding as to the processes and how dangerous they can be. The remainder of this discussion will consider these critical gaps in understanding and the role scientists can play in closing them.

Perception of hazards, exposure and risk

Perception of hazards, exposure and risk determines how a population will respond in a volcano-related emergency; it is conditioned by the history, culture, level of education, state of economic development and social structure of that population (United Nations, 1985). Furthermore, public response to warnings in an emergency is a complicated sociological and psychological process that involves the attributes of the sender of the warning message, the attributes of the warning message itself, and the attributes and experience of the receivers (Sorensen and Mileti, 1987). It may be that neither the people living at the base of a volcano nor the responsible government officials are able to conceive of a catastrophic debris avalanche or large debris flow if they have never seen or heard of such events. If they have experienced a flow event previously, it may have been small enough to preclude any worry of future events as being dangerous. In 1985, misperceptions of a hydrologic hazard (volcanic debris flows), the exposure of their town, and the true danger such flows presented played a large part in the tragedy in Armero, Colombia (Bruno Podesta, Grupo de Estudios para el Desarrollo, Lima, oral communication, 1987).

Other important factors will determine how a population responds to a hazard, both in the long term (planning) and short term (emergency response to warning). There may be physical, emotional, economic, or political constraints placed on the options available for response, and there may be complicated interactive links between factors. But a correct perception of the nature of the hazards, the degree of vulnerability, and the probabilities of

TABLE 3 *Strategies for response to potential hydrologic hazards at volcanoes*

Strategy	Advantages/Disadvantages
Accept risk; do nothing.	Cheapest strategy but offers no protection; may be appropriate if risk is sufficiently low.
Provide financial incentives or disincentives for removing existing development and preventing/regulating new development in hazardous areas.	Requires that governmental agencies have a good understanding of the hazards and risks. May be very costly.
Build protective structures for existing development.	Offers protection from events only up to a certain 'design' magnitude; will not protect against larger events. May foster false sense of security. Cost depends on scale.
Implement real-time warning systems and establish nearby refuges for residents.	Depends on reliability of warning system, which could fail in real event or give false alarms that would lead people to ignore it. 'Low-tech' varieties relatively cheap and may be more reliable in harsh environmental conditions. Property not protected.
General evacuation.	Costly. Requires relatively long time to accomplish (too long for debris avalanches and large debris flows close to volcano). Difficult to decide when to order evacuation.
Exclusionary zoning.	Only guaranteed effective strategy if zone is correctly delineated. May be politically unacceptable. May be very costly.

magnitude and frequency of occurrence is the vital first step to an adequate response. Scientists should be best able to improve faulty perceptions.

The scientists' roles: are new ones needed?

Programmes designed to reduce the impact of natural hazard on humans and their property have five basic components (W.J. Kockelman, US Geological Survey, written communication, 1988), whether the hazards are volcanically triggered sediment-water flows or any other natural hazard. Furthermore, the 'all-hazards approach' (i.e. a scheme independent of the type of hazard) is considered by some to be the best approach to disaster planning and preparedness (Wenger, 1988). The five basic components of an effective response to a potential hazard are:

(1) *Investigation*, carrying out those scientific and engineering studies necessary to understand the characteristics of the hazardous physical processes, to predict the areas of probable impact, to determine the probabilities of occurrence, and to assess the probable response of structures and other cultural facilities to the processes.

(2) *Translation*, putting the results of scientific and engineering studies into reports and onto maps so that the nature of the hazard, the areas at risk, and the degrees of risk can be understood by nontechnical users.

(3) *Transfer*, conveying the above information to the appropriate civil authorities and to the public (directly and through the new media), and assisting and encouraging them in the use of this information.

(4) *Implementation*, selecting, planning, carrying out, and maintaining (over the long term) the most appropriate mitigative measures to prepare for a disaster, including formulation and periodic rehearsals of a disaster response plan, as well as making preparations for emergency response during an event and for rescue, recovery, and reconstruction (or relocation) after an event.

(5) *Review*, reviewing the effectiveness of the disaster mitigation techniques after they have been in use for a requisite time and revising as necessary.

In the past, scientists typically have had responsibility for the first two steps in this sequence, passing the studies and hazard maps on to civil defence and other governmental officials for them to complete the sequence of steps. The example of the Ruiz disaster proves that this is not enough. Steps 1 and 2 (including the publication of accurate hazard maps) were competently done by the scientific community in Colombia for Nevado del Ruiz prior to the devastating debris flows of 13 November 1985 (Herd *et al.*, 1986), but significant failures occurred in the remaining steps, and a catastrophe resulted.

Scientists need to be intimately involved in step 3, the delivery of information and the assistance and encouragement to use it effectively. A big advantage is that in most cases, scientists are viewed as a credible source of

229

information and have a good understanding of the hazardous processes. But *how* that information is transferred is most important. To be effecively transferred, the information must be understandable (without jargon), specific, complete, and be a consistent consensus opinion of all the scientists involved (United Nations, 1985; Sorensen and Mileti, 1987). Behaviour research has shown that the public's first response to a warning message is to try to confirm the information by another credible source before taking action (Sorensen and Mileti, 1987). It is therefore critical that the scientific community and all other local, state and national officials involved deliver a single message, not several conflicting ones, so that the confirmation process can be quickly completed. The lack of consistent, credible information at the time of the emergency is cited as a major contributing factor to the Armero disaster (Podesta and Olson, 1988). Ways of encouraging more participation by scientists in this transfer process need to be found.

Scientists also need to play a major advisory role in step 4, particularly in the selection and planning stages. Major changes may be required in leglislation, regulations, design criteria, financial incentives, and public and corporate policies. The inertia resisting such changes may be large, yet only the scientific community has the ability to credibly demonstrate what the consequences of inaction might be. Scientists also are in the best position to evaluate the potential effectiveness of various response plans. Again, the scientific community needs to reach consensus and to speak with one voice so that extreme views and untested hypotheses do not hold sway, as happened in Peru a few years ago with the Brady earthquake prediction (Podesta and Olson, 1988). Forums need to be established for the quick and efficient reaching of scientific consensus in times of emergency and for the regular and ongoing interaction of the scientific community with disaster preparedness officials.

Scientists also must be involved in step 5, particularly in encouraging the need for review and reassessment as natural conditions change, as cultural patterns change, and (especially) as new personnel assume positions of responsibility in disaster preparedness. Population pressure may lead to residential development in areas designated as hazardous. Countering such development on the grounds of potential hazards may be difficult if pro-development pressure is strong or if, as individual officials leave an agency, the agency 'forgets' about the hazards.

Conclusion

The potential hydrologic consequences of volcanic eruptions include debris avalanches, debris flows, and various categories of floods, and all of them can cause catastrophic loss of life and property. Adequate response to an impending hydrologic event, such as one of these, requires the effective completion of five basic hazard response steps: investigation, translation of information, transfer of information, implementation of the programme, and review of the programme. Hydrological hazards from volcanoes have certain characteristics that distinguish them from other natural hazards, but the process of anticipating and avoiding disasters is essentially the same for all types of hazards.

Deciding how and when to respond will not be done by scientists but by the population at risk and by governmental officials at local, regional and national levels. Timely, effective response to any potential hazard requires the participation of the scientific community at each of the five steps in the hazard response and at each of the governmental levels, so that misperceptions about the hazards and the dangers they pose can be eliminated. Such misperceptions may be a fundamental cause for lack of adequate disaster preparation. Additional and more effective ways need to be found to (1) bring scientists together to reach consensus about impending disasters, and (2) build trusting, ongoing, professional relationships between scientists and government officials responsible for disaster preparedness. Future programmes under the International Decade for Natural Disaster Reduction could aid in achieving these goals.

REFERENCES

AGUILA, L.G., NEWHALL, C.G., MILLER, C.D. and LISTANCO, E.L. (1986) Reconnaisance geology of a large debris avalanche from Iriga volcano, Philippines. *Philippine J Volcanol* **3**: 54–72

ARAMAKI, S. (1956) The 1783 activity of Asama volcano. Part I. *Japanese J Geol Geogr* **27**: 189–229

ASHIDA, K. and EGASHIRA, S. (1986) Running-out processes of the debris associated with the Ontake landslide. *Natural Disaster Sci* **8**: 63–79

BAKER, P.E. (1969) A volcano erupts beneath the Antartic ice. *Geograph Mag* **41**: 115–126

BAKER, P.E., DAVIES, T.G. and ROOBOL, M.J. (1969) Volcanic activity at Deception Island in 1967 and 1969. *Nature* **224**: 553–560

BJORNSSON, H. (1974) Explanation of jokulhlaups from Grimsvotn, Vatnajokull, Iceland. *Jokull* **24**: 1–14

BJORNSSON, H. (1975) Subglacial water reservoirs, jokulhlaups, and volcanic eruptions. *Jokull* **25**: 1–14

BJORNSSON, H. and KRISTMANNSDOTTIR, H. (1984) The Grimsvotn geothermal area, Vatnajokull, Iceland. *Jokull* **34**: 25–50

BLONG, R.J. and JOHNSON, R.W. (1986) Geological hazards in the southwest Pacific and southeast Asian region: identification, assessment, and impact. *BMR J Australian Geol Geophys* **10**: 1–15

CHILDERS, D. and CARPENTER, P.J. (1985) A warning system for hazards resulting from breaches of lake blockage. *Proc Int Symp Erosion, Debris Flow, and Disaster Prevention.* Tsukuba, Japan, 493–498

COLLINS, B.D. and DUNNE, T. (1986) Erosion of tephra from the 1980 eruption of Mount St. Helens. *Geol Soc Am Bull* **97**: 896–905

COSTA, J.E. and SCHUSTER, R.L. (1988) The formation and failure of natural dams. *Geol Soc Am Bull* **100**: 1054–1068

CRANDELL, D.R. (1971) Postglacial lahars from Mount Rainier Volcano, Washington. *U.S. Geol Survey Prof Paper* **677**: 73 pp

CRANDELL, D.R. (1988) Gigantic debris avalanche of Pleistocene age from ancestral Mount Shasta volcano, California, and debris-avalanche hazard zonation. *U.S. Geol Survey Bull* 1861: 32 pp

CRANDELL, D.R., BOOTH, B., KUSUMADINATA, K., SHIMOZURU, D., WALKER, G.P.L., and WESTERCAMP, D. (1984) *Source book for volcanic hazards zonation.* UNESCO, Paris, 97 pp

CUMMANS, J. (1981) Chronology of mudflows in the South Fork and North Fork Toutle River following the May 18 eruption. In: Lipman, P.W. and Mullineaux, D.R. (Eds). The 1980 Eruptions of Mount St Helens, Washington. *U.S. Geol Survey Prof Paper* **1250**: 479–486

DUNNE, T. and LEOPOLD, L.B. (1981) *Flood and sedimentation hazards in the Toutle and Cowlitz River system as a result of the Mt. St. Helens eruption.* U.S. Federal Emergency Management Agency, Region 10, 159 pp

FAIRCHILD, L.H. (1985) Lahars at Mount St. Helens, Washington. Ph.D. Dissertation, University of Washington, Seattle, Washington, 374 pp

GALLINO, G.L. and PIERSON, T.C. (1985) Polallie Creek debris flow and subsequent dam-break flood of 1980, East Fork Hood River Basin, Oregon. *U.S. Geol Survey Water-Supply Paper* **2273**: 22 pp

GAGOSHIDZE, M.S. (1969) Mud flows and their control. *Soviet Hydrol Selected Papers,* **4**: 410–422

GONZALES-FERRAN, O. (1985) Description of volcanic events – Chile (Villarrica volcano). *SEAN Bulletin* **10**: 3

HEALY, J., LLOYD, E.F., RISHWORTH, D.E.H., WOOD, C.P., GLOVER, R.B., DIBBLE, R.R., TRAILL, W.E., MAZEY, J.W., COLLINS, C.M. and BURSTALL, P.J. (1978) The eruption of Ruapehu, New Zealand, on June 22 1969. *DSIR Bull* **224**: 80 pp

HERD, D.G. and COMITE DE ESTUDIOS VULCANOLOGICOS (1986) The 1985 Ruiz volcano disaster. EOS, *Trans Am Geophys Union* **67**: 457–460

HEWSON, C.A.Y. and LATTER, J.H. (1976) *Lahar warning system, Ruapehu.* New Zealand Dept. Sci. Indus. Res., Geophysics Div., unpublished report

HUNGR, O., MORGAN, G.C. and KELLERHALS, R. (1984) Quantitative analysis of debris torrent hazards for design of remedial measures. *Canadian Geotech J* **21**: 663–677

INOKUCHI, T. (1985) The Ontake rock slide and debris avalanche caused by the Naganoken-Seibu earthquake, 1984. In: *Proc. IVth Int Conf Field Workshop on Landslides,* Tokyo, 329–338

JANDA, R.J., SCOTT, K.M. and MARTINSON, H.A. (1981) Lahar movement, effects, and deposits. In: Lipman, P.W. and Mullineaux, D.R. (Eds). The 1980 Eruptions of Mount St. Helens, Washington. *U.S. Geol Survey Prof Paper* **1250**: 461–478

JANDA, R.J., BANKS, N.G., PIERSON, T.C., CALVACHE, M.L. and THOURET, J.C. (1986) Interaction between ice and pyroclastic sediments erupted during the 13 November 1985 eruption of Nevado del Ruiz, Colombia (abst.). EOS, *Trans Am Geophy Union* **67**: 406

JENNINGS, M.E., SCHNEIDER, V.R. and SMITH, P.E. (1981) Emergency assessment of Mount St. Helens post-eruption flood hazards, Toutle and Cowlitz Rivers, Washington. *U.S. Geol Survey Circular* **850-I**, 7 pp

JONSSON, J. (1982) Notes on Katla volcanoglacial debris flows. *Jokull* **32**: 61–68

KOCKELMAN, W.J. (1984) Reducing losses from earthquakes through personal preparedness. *U.S. Geol Survey Open-File Report* **84–765**, 13 pp

KOCKELMAN, W.J. (1986) Some techniques for reducing landslide hazards. *Bull Assoc Eng Geol* **23**: 29–52

KONINKLIJK MAGNETISCH EN METEOROLOGISCH OBSERVATIONS TE BATAVIA (1929) Volkanische erschijnselen en aardbevingen in den Oost-Indischen Archipel waargenomen gedurende het jaar 1928. *Nat Tijdschr v Ned Ind* **89**: 173–209

LEHRE, A.K., COLLINS, B.D. and DUNNE, T. (1983) Post-eruption sediment budget for the North Fork Toutle River drainage, June 1980-June 1981. In: Okuda, S., Netto, A. and Slaymaker, O. (Eds). *Extreme Land Forming Events: Zeitschrift fuer Geomorphologie, Suppl Bd*, **46**: 143–163

LOMBARD, R.E., MILES, M.B., NELSON, L.M., KRESH, D.L. and CARPENTER, P.J. (1981) The impact of mudflows of May 18 on the lower Toutle and Cowlitz Rivers. In: Lipman, P.W. and Mullineaux, D.R. (Eds). The 1980 Eruptions of Mount St. Helens, Washington. *U.S. Geol Survey Prof Paper* **1250**: 693–699

LOWE, D.R., WILLIAMS, S.N., LEIGH, H., CONNOR, C.B., GEMMELL, J.B. and STOIBER, R.E. (1986) Lahars initiated by the 13 November 1985 eruption of Nevado del Ruiz, Colombia. *Nature* **324**: 51–53

MAJOR, J.J. and NEWHALL, C.G. (1989) Snow and ice perturbation during historical volcanic eruptions and the formation of lahars and floods – A global review. *Bull Volcanol*, in press

MAJOR, J.J. and VOIGHT, B. (1986) Sedimentology and clast orientations of the 18 May 1980 southwest-flank lahars, Mount St. Helens, Washington. *J Sed Petrology* **56**: 691–705

MARTIN, D.C., PITEAU, D.R., PEARCE, R.A. and HAWLEY, P.M. (1984) Remedial measure for debris flows at the Agassiz Mountain Institution, British Colombia. *Canadian Geotech J* **21**: 505–517

MEYER, D.F. and JANDA, R.J. (1986) Sedimentation downstream from the 18 May 1980 North Fork Toutle River debris avalanche deposit, Mount St. Helens, Washington. *Mount St. Helens: Five Years Later*, Eastern Washington University Press, 68–86

MOJICA, J., COLMENARES, F., VILLARROEL, C., MACIA, C. and MORENO, M. (1985) Caracteristicas del flujo de lodo ocurrido el 13 de noviembre de 1985 en el valle de Armero (Tolima, Colombia). Historia y comentarios de los flujos de 1595 y 1845. *Geologica Colombiana* **14**: 107–140

MOORE, J.G. and MELSON, W.G. (1969) Nuees ardentes of the 1968 eruption of Mayon volcano, Philippines. *Bull Volcanol* **33**: 600–620

MORENO ROA, H., GONZALES-FERRAN, O. and RIFFO, P. (1981) Llaima volcano. Annual Rpt. of World Volc. Eruptions in 1979. *Bull Volc Eruptions* **19**: 76

NAKAMURA, S. (1926) On the velocity of recent mud-flows in Japan. *Proc 3rd Pan-Pacific Congress*, Tokyo, 788–800

NARANJO, J.L., SIGURDSSON, H., CAREY, S.N. and FRITZ, W.J. (1986) Eruption of the Nevado del Ruiz volcano, Colombia, on 13 November 1985: tephra fall and lahars. *Science* **233**: 961–963

NASMITH, H.W. and MERCER, A.G. (1979) Design of dykes to protect against debris flows at Port Alice, British Colombia. *Canadian Geotech J* **16**: 748–757

NEALL, V.E. (1976) Lahars as major geological hazards. *Bull Int Assoc Eng Geol* **14**: 233–240

OINOUYE, Y. (1917) A few interesting phenomena on the eruption of Usu. *J Geol* **25**: 258–288

OKKERMAN, J.A., GERONIMO, S.G., UMBAL, J.V. and PALAD, J.N. (1985) Study into the mechanisms of debris flow during/after the 1984 eruption of Mayon volcano. *Philippine J Volcanol* **2**: 94–142

OMORI, F. (1911) The Usu-san eruption and earthquake and elevation phenomena. *Bull Imperial Earthquake Investigative Committee* **5(I)**: 1–38

O'SHEA, B.E. (1954) Ruapehu and the Tangiwai Disaster. *New Zealand J Sci Tech* **B36**: 174–189

PATERSON, B.R. (1980) The hazard of lahars to the Tongariro power development. *Proc 3rd Austral-NZ Geomech Conf, N.Z.I.E. Proc Tech Groups 6/1(G):* Christchurch, N.Z., 2–7 – 2–14

PIERSON, T.C. (1985) Initiation and flow behaviour of the 1980 Pine Creek and Muddy River lahars, Mount St. Helens, Washington. *Bull Geol Soc Am* **96**: 1056–1069

PIERSON, T.C. and SCOTT, K.M. (1985) Downstream dilution of a lahar: transition from debris flow to hyperconcentrated streamflow. *Water Resources Res* **21**: 1511–1524

PIERSON, T.C., JANDA, R.J., THOURET, J.C. and CALVACHE, M.L. (1986) Initiation, bulking, and channel dynamics of catastrophic debris flows on 13 November 1985 at Nevado del Ruiz, Colombia. *EOS, Trans Am Geophys Union* **67**: 406

PIERSON, T.C. and COSTA, J.E. (1987) A rheologic classification of subaerial sediment-water flows. *Geol Soc Am Rev Eng Geol* **VII**: 1–12

PODESTA, B. and OLSON, R.S. (1988) Science and the state in Latin America: Decisionmaking in uncertainty. In: Comfort, L.K. (Ed.) *Managing Disaster, Strategies and Policy Perspectives:* Duke University Press, Durham and London, pp 296–312

RODOLFO, K.S., ARGUDEN, A.T., SOLIDUM, R.A. and UMBAL, J.V. (1989) 17 October 1985 rain-induced lahar on Mayon Volcano, Philippines. *Int Assoc Eng Geol*, in press

RICHARDSON, D. (1968) Glacier outburst floods in the Pacific Northwest. *U.S. Geol Survey Prof Paper* **600-D**: D79–D86

SCHUSTER, R.L. and CRANDELL, D.R. (1984) Catastrophic debris avalanches from volcanoes. *Proc IVth Internat Conf on Landslides*, Toronto, 567–572

SCOTT, K.M. (1988) Origins, behavior, and sedimentology of lahars and lahar-runout flows in the Toutle-Cowlitz River system. *U.S. Geol Survey Prof Paper* **1447-A**: 74 pp

SIEBERT, L. (1984) Large volcanic debris avalanches: characteristics of source areas, deposits and associated eruptions. *J Volcanol Geotherm Res* **22**: 163–197

SIEBERT, L., GLICKEN, H. and UI, T. (1987) Volcanic hazards from Bezymianny- and Bandai-type eruptions. *Bull Volcanol* **49**: 435–459

SILVA, L., COCHEME, J.J., CANAL, R., DUFFIELD, W.A. and TILLING, R.I. (1982) The March-April 1982 eruptions of Chichonal Volcano, Chiapas, Mexico – Preliminary observations (abst.). *EOS Trans Am Geophys Union* **63**: 1126

SMART, G.M. (1981) Volcanic debris control, Gunung Kelud, East Java. Proc. of Conf. on Erosion and Sediment Transport in Pacific Rim Steeplands, Christchurch, N.Z. *Int Assoc Hydrol Sci Publ* **132**: 604–623

SORENSEN, J.H. and MILETI, D. (1987) Public warning needs. In: Gori, P.L. and Hays, W.W. (Eds). Proceedings of Conference XL: A Workshop on 'The U.S. Geological Survey's Role in Hazards Warnings.' *U.S. Geol Survey Open File Report* **87-269**: 9–75

SUDRADJAT, A. and SISWOWIDJOYO, S. (1981) Descriptions of volcanic eruptions – Indonesia. *Bull Volcanic Eruptions* **19**: 33–36

SUMARYONO, A. and KONDO, K. (1985) Development programme of volcanic mud-flow forecasting and warning system in Indonesia. *Proc Internat Symp on Erosion, Debris Flow, and Disaster Prevention:* Tsukuba, Japan, 481–486

SURYO, I. and CLARKE, M.C.G. (1985) The occurrence and mitigation of volcanic hazards in Indonesia and exemplified at the Mount Merapi, Mount Kelut and Mount Galunggung volcanoes. *Q J Eng Geol London* **18**: 79–98

THORARINSSON, S. (1953) Some new aspects of the Grimsvotn problem. *J Glaciol* **2**: 269–275

THORARINSSON, S. (1975) Katla og annall kotlugosa. *Ferdafelag Islands Arbok*, 125–149

THOURET, J.C., JANDA, R.J., PIERSON, T.C., CALVACHE, M.L. and CENDRERO, A. (1987) L'eruption du 13 novembre 1985 au Nevado El Ruiz (Cordillere Centrale, Colombie): interactions entre le dynamisme eruptif, la fusion glaciaire et la genese d'ecoulements volcano-glaciaires. *C R Acad Sci Paris* **305**: 505–509

TOMBLIN, J. (1982) Managing volcanic emergencies. *UNDRO News*, Jan. 1982, 4–10

UMBAL, J.V. (1989) Recent lahars of Mayon Volcano. *Philippine J Volcanol* **3**: in press

UNITED NATIONS (1977) *Disaster prevention and mitigation, Vol. 1 – Volcanological aspects.* Office of the UN Disaster Relief Co-ordinator (UNDRO), Geneva, 68 pp

UNITED NATIONS (1985) *Volcanic emergency management.* Office of the UN Disaster Relief Co-ordinator (UNDRO), Geneva, 86 pp

VANONI, V.A. (1982) Comments on sedimentation problems resulting from the eruption of Mt. St. Helens. In: Washington Water Research Center, Proc. from the conference, Mount St. Helens – Effects on Water Resources: Pullman, Wash., *Wash Water Res Center Rep* **41**: 395–403

VAN PADANG, M.N. (1960) Measures taken by the authorities of the Vulcanological Survey to safeguard the population from the consequences of volcanic outbursts. *Bull Volcanol ser. 2,* **23**: 181–193

VARNES, D.J. (1978) Slope movement types and processes. In: Schuster, R.L. and Krizek, R.J. (Eds). *Landslides: Analysis and Control.* Washington, D.C., National Academy of Sciences, Transportation Research Board Special Report 176, 11–33

VOIGHT, B., GLICKEN, H., JANDA, R.J. and DOUGLASS, P.M. (1981) Catastrophic rockslide avalanche of May 18. In: Lipman, P.W. and Mullineaux, D.R. (Eds). The 1980 Eruptions of Mount St. Helens, Washington. *U.S. Geol Survey Prof Paper* **1250**: 347–377

VOIGHT, B., JANDA, R.J., GLICKEN, H. and DOUGLASS, P.M. (1983) Nature and mechanics of the Mount St. Helens rockslide-avalanche of 18 May 1980. *Geotechnique* **33**: 243–273

WAITT, R.B., PIERSON, T.C., MACLEOD, N.S., JANDA, R.J., VOIGHT, B. and HOLCOMB, R.T. (1983) Eruption-triggered avalanche, flood, and lahar at Mount St. Helens – effects of winter snowpack. *Science* **221**: 1394–1396

WALDRON, H.H. (1967) Debris flow and erosion control problems caused by the ash eruptions of Irazu volcano, Costa Rica. *U.S. Geol Survey Bull* **1241-I**: 37pp

WATANABE, M. and IKEYA, H. (1980) Research and countermeasure for sediment disasters in the surrounding area of active volcano. *Symp for the Commission on Field Experiments in Geomorphology,* IGU, Japan, 1–6

WATANABE, M. and IKEYA, H. (1981) Investigation and analysis of volcanic mud flows on Mt. Sakurajima, Japan. Proceedings of Symposium on Erosion and Sediment Transport Measurement, Florence, Italy: *Int Assoc Hydrol Sci* **133**: 245–256

WENGER, D. (1988) Volcanic disaster prevention, warning, evacuation and rescue systems (abst.). *Proceedings Kagoshima International Conference on Volcanoes,* Kagoshima, Japan, 456

Volcanoes and Hydrology in Indonesia

ALI HAMZAH LUBIS

Institute of Hydraulic Engineering,
Bandung, Indonesia

Introduction

INDONESIA consists of a chain of islands, extending from 95° to 140° West longitude and is divided into two parts by the Equator (Figure 1). In total there are more than 10,000 islands of which 5,000 are inhabited. The population is 175 million people with annual growth of 2%, almost half of them on the island of Java. Other islands with significant population numbers include Sumatera, Sulawesi, Kalimantan and Irian Jaya.

Many of the islands have active volcanoes, one of the most famous was Krakatoa which erupted in 1883 causing severe loss of life and a dust cloud which affected the transmission of sunlight on a global scale and for a period of several years. Other less spectacular eruptions occur more frequently and almost every year there are a number of eruptions.

Mount Galunggung – A Case Study

The most recent major eruption in April 1982 was that of Mount Galunggung, in West Java (Figure 2). The eruption caused great damage to the surrounding area, among other things to the animal and plant life, plantations, irrigation systems, construction, etc. There have been four recorded eruptions of Mount Galunggung: in 1882, 1894, 1918 and 1982. The eruption in 1982 lasted more than 9 months.

Mount Galunggung is one of the volcanoes in the mountain chain which stretches from the northernmost end of Sumatera through Java to the islands of Bali, Lombok and Sumbawa. It is the catchment divide between the River Cimanuk, which flows north to the Java Sea, and the rivers Ciwulan and Citanduy which flow southwards to the Indian Ocean. The main eruption was

237

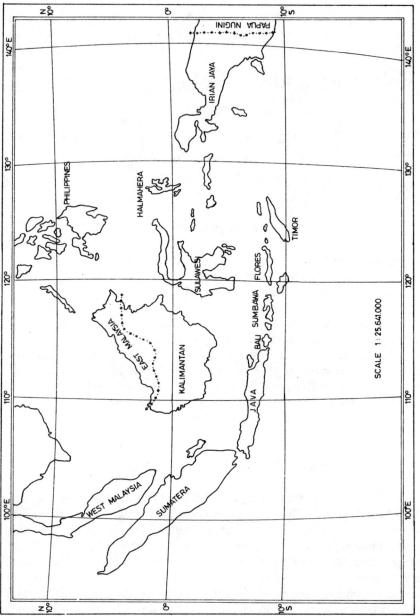

FIGURE 1 *Map of Indonesia*

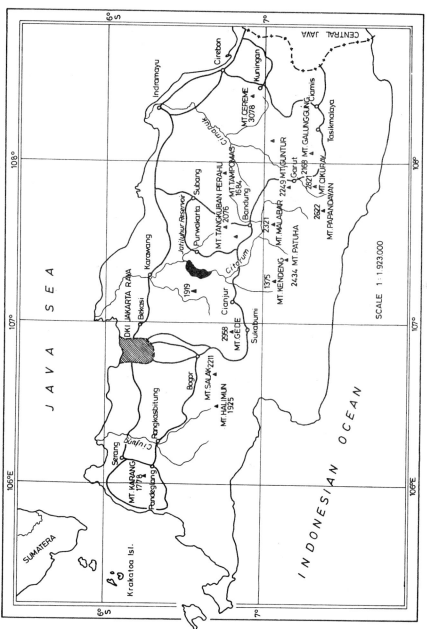

FIGURE 2 *West Java*

on the slopes of the mountain facing these south-flowing catchments. The quantity of dust released was immense. In villages near to the volcano, deposits up to 3 m deep were formed and the smoke and dust plume rose to above 10,000 m, so that air traffic had to be detoured either to the north or south. In parts of the island the sun was barely visible on the day after the eruption.

The elevation of Mount Galunggung is 2,168 m above mean sea level, and it is located in an area with temperatures ranging between 24°C and 25°C, with an annual runoff of about 3,200 mm. The horseshoe-shaped surrounding area is a rice growing area of great potential with semi-technical irrigation and dense population: around 720 people per km^2. Another potential growth area is the fish-farm enterprises which make use of the traditional system. Both of these require regular fresh water for growth of rice plants and fish.

Sediment material ejected from the last volcanic eruption was estimated at 370 million m^3. The runoff of sediment has continued since then, with sediment deposited on the mountainside being washed down during the rainy season and entering the river systems. This affected the irrigation channels with a damaging effect on agriculture in addition to the fact that topsoil was buried below a layer of dust. To reduce the damaging effects of silt entering the irrigation channels and river systems, the first solution tried was to construct check dams and sand pockets to detain the sediment but continual erosion from the unvegetated slopes has rapidly filled them up.

Table 1 shows the capacity of the check dams (in millions of m^3) and the available capacity.

It is, however, estimated that the quantity of silt remaining on the slopes is 10 million m^3 of which about 6 million m^3 is liable to enter the dams and river systems. Little of this will be held back by the check dams as they are nearing full capacity and their trap efficiency is low. Another solution being examined at the moment is to construct a diversion channel on the slopes of the mountain which will carry much of the runoff and therefore silt from the upper reaches of the Ciloseh/Citanduy river system into the Ciwulan river system. The Citanduy is more important from an irrigation point of view. Although there will be some loss of available water the benefits from lower siltation will be considerable.

The sand pockets and check dam have in principle the same function, i.e. to detain sediment transport and prevent it from entering areas in need of protection. Detention of sediment is only of a temporary nature because, as explained earlier, both structures will be full of sediment in no time. Therefore, the sediment deposit detained there has to be extracted periodically and transported by vehicle to a safe place and from there onwards to areas needing it for construction material. The difference between the sand pocket and check dam lies only in their location. The sand pocket is not located in the river channel but on an area estimated to be passed by the sediment flow. It is a wall extending in the low land, whereas the check dam is a dam structure but

TABLE 1
Capacity of check dams

Name	Total capacity (10^6 m^3)	Remaining capacity (10^6 m^3)
Ciponyo I (Inner)	2.5	0.3
Ciponyo I	6.7	0.7
Ciponyo II	4.7	1.6
Cimampang	1.7	0.1
Negla	3.4	1.2

with a different function from that of an ordinary dam. The check dam only functions to trap the sediment transport.

The volume of sediment in the sand pockets and check dams has reached its maximum limit because this prevention policy is being implemented. Extensive mineral workings have been developed since it was discovered that the volcanic silt was suitable for concrete. To export this, a plant for washing the silt and loading it onto railway trains has been installed and the railway network has been extended to the extraction site.

One particular problem is that a volcanic lake has been formed at the top of Mount Galunggung, upstream of the town of Tasikmalaya with its population of about half a million, and one of the areas where sediment is abstracted. This is about 500 m in diameter and several tens of metres deep, and stores around 7.5×10^6 m^3 of water. Water accumulated in the crater lake after eruption has shown a tendency to rise and the crater lake wall is unstable. Hence there are fears that a sudden collapse could lead to catastrophic flooding with loss of life and severe damage to plant installed for sediment extraction.

It is considered that a telemetry scheme to monitor the lake would be justified. This would be an event-based system which would have two level sensors in the lake itself and a few downstream on the rivers. These outstations would normally send a message every 12 hours or whenever the level varied by more than 5 cm. It is likely that dam collapse would be initiated either by piping, leading to internal collapse of the dam, or by overflow. With such a telemetry system a more rapid than usual drop in level could be a signal that piping losses were increasing; similarly, the level at which overtopping occurs would be known and any high levels could be closely monitored. If having once reached the overtopping level the lake level started dropping rapidly, then this would indicate erosion of the bank by the discharge. In either of the above cases appropriate actions could be taken to minimise loss of life and damage to property.

In the Cimanuk a telemetry project funded by the UNDP, and for which the WMO is the executing agency, is currently under way. It is based at the Institute of Hydraulic Engineering, Bandung. The main aim of this pilot project is to improve water management of the Cimanuk basin and to develop techniques of general application in Indonesia. One aspect which will be receiving attention is the use of turbidity sensors to estimate silt load. Traditional methods of measuring sediment rely on point measurements and cannot therefore trace the build-up of sediment load following volcanic action. By measuring turbidity on a continuous basis and relating it to sediment load a more complete picture of the variation of suspended sediment can be obtained. A study of the results of sampling in the river before and after the volcano had erupted showed that at sediment loads of less than about 700 parts per million, equivalent to a flow of 100 m^3/s, there was little change in the flow/suspended solids curves but there was a marked difference at higher flows with twice as much sediment being recorded at higher flows.

In the lower parts of the Cimanuk basin is an irrigation scheme with a potential command area of 250,000 hectares, although only about half of this is currently developed. By monitoring the sediment in the river whether from volcanoes or erosion and controlling the diversion of water for irrigation the costly damaging effects of siltation of the irrigation channels can be reduced.

One problem which still remains to be solved is that of identifying the source of river-borne silt. Chemical analysis is not able to help as the parent material of the soils is of the same nature as the highest rates of erosion in the world but this may be misleading. There is little evidence of the fan-shaped erosion gullies which normally characterize very high rates of erosion and it is therefore possible that the volcanoes are a major contributing factor.

Acknowledgement

The views expressed in this paper are those of the author and do not necessarily represent those of the Institute of Hydraulic Engineering. The author wishes to thank the Director of the IHE for permission to publish this paper. He also gratefully thanks his colleagues at the Institute and in the hydrology project for their suggestions and support.

Non-Meteorological Flood Disasters in Chile

HUMBERTO PEÑA[1] and WULF KLOHN[2]

[1]*Hydrology Department, Direccion General de Aguas, Santiago, Chile,*
[2]*Hydrology and Water Resources Department, World Meteorological Organization, Geneva, Switzerland*

Introduction

CONTINENTAL Chile, situated on the South American continent to the West of the Andes Cordillera, spans some 4,500 km between 68° and 75° west longitude and 17° and 56° south latitude, with an average width of only about 190 km (Figure 1). The climate is arid and semi-arid in the north, temperate of Mediterranean type in the centre, and humid temperate in the south. This territory has several characteristics that favour the occurrence of non-meteorological floods, including:

(a) glaciers that cover a total area of some 20,000 km². (These glaciers exist in the Cordillera under varied climate conditions from North to South. In the arid northern region, the glaciers, situated at high altitude, are small and show almost no activity. However, in the southern region there are large very active ice fields that reach all the way to the ocean.);

(b) mountains that cover roughly 80% of the territory, including some 300 peaks of more than 5,000 m;

(c) intense volcanic and seismic activity due to the position of the country on the active fringe of tectonic plates. (Nearly 60% of the territory is covered by Cenozoic and modern rocks whose origin is related to vulcanism, and there are nearly 50 active volcanoes, about 10% of the world's total. Of these, 24 are situated between 33° and 41° south latitude, which is the more densely populated part of the country. Seismic activity is reflected in the occurrence of one major earthquake at any place every 80 to 100 years on average.).

FIGURE 1 *Location map*

Floods caused by non-meterological events have mostly happened in sparsely populated and undeveloped regions of the country. However, more recently such events have taken both lives and property, and hydrologists as well as the public have become aware of the need to understand these phenomena. The results of recent research to determine the magnitude, origin and propagation of non-meteorological floods observed in Chile during the past 40 years show the need to improve the methods applied to hazard assessment and risk evaluation.

244

Information Available

Generally, the information available on non-meteorological floods is incomplete, nonstandard and dispersed in unpublished reports of different agencies and is therefore difficult to access. Systematic studies are scarce and apply only to the more recent events.

In this paper, the results of studies carried out on various types of floods that were not triggered by meteorological events are discussed in an attempt to present a systematic overview of these phenomena. The sites of eight flood events of non-meteorological origin are shown in Figure 1. Table 1 contains a summary of data concerning the floods selected, including background information on losses, the length of the river reach where discharge was larger than 100-year meteorological flood (Q_{100}), volume, peak discharge, duration of the flood, and the Q_{100} value corresponding to the initial section of the flood. It can be seen that the peak discharges are quite different from each other but always much larger than the respective Q_{100}.

It is worth noting that for the more recent floods discussed here, the records from the national hydrometric network provided valuable data that were not available for earlier events.

Floods Generated by Glacier Outburst

There are cases where the sudden release of water by glaciers is repeated regularly and is not therefore an unforeseen disaster. Only cases of unforeseen sudden outburst of water retained by glaciers are discussed here. Such floods have been observed in Chile under varying conditions. They can be classified according to the process responsible for the accumulation of the mass of water and according to the emptying process of the accumulated water. The following accumulation processes have occurred:

(a) An advancing glacier blocks a tributary thus creating a lake. This situation was at the origin of the 2 February 1954 flood of the Rio Olivares (Table 1, case 2). The Rio Olivares is situated in the Andes of Central Chile and is the natural drainage of four fairly large glaciers covering a total area of 75 km². One of the glaciers had advanced five years earlier, blocking the drainage of the other three and impounding a small lake of some $0.4 \times 10^6 \, m^3$ which eventually overtopped the ice barrier and emptied in a short period of time. (Chilectra, 1954; Lliboutry, 1955).

(b) Accumulation of water owing to obstruction of the normal drainage through the ice. This is the case with the floods of the Rio Paine fed by the Dickson glacier, at the southern end of the country (Table 1, case 8), and of the Rio Cachapoal fed by the glacier of the same name 100 km south of Santiago (Table 1, case 4).

TABLE 1

Summary data of eight non-meteorological floods

River	Date	Initial point	Altitude (m)	Type of phenomenon	D* (km)	End point**	Losses	Volume (10^6 m³)	Q_{max} (m³/s)	Time	Q_{100} (m³/s)	Q_{max}/Q_{100}
Manflas	14.05.85	28° 33' S 69° 43' W	5,200	Emptying of subglacial lake/structural failure	105	reservoir[1]	small[1]	5	11,000	15 min	< 1	⩾ 10
Olivares	26.02.54	33° 10' S 70° 07' W	2,800	Emptying of lake/advancing glacier	50	main river	small[2]	0.4	400	30 min	100	4
Colorado	29.11.87	33° 19' S 70° 02' W	2,950	Landslide followed by flow of detritus	59	main river	large[3]	7	> 10,000	10 min	20	⩾ 10
Cachapoal	31.01.81–17.02.81[4]	34° 22' S 70° 07' W	2,270	Emptying of lakes retained by glacier	22	main river	small[2]	1.5–2	150	10 h	60	2.5
Volcan Villarrica	29.12.71	39° 25' S 71° 57' W	2,800	Lahars	15	lakes	large[6]	15[11]	3,500[10,11]	4 h	150	⩾ 10
San Pedro	26.07.60	39° 45' S 72° 30' W	120	Emptying of lake retained by landslide	90	ocean	small[7]	2,000	7,450	10 days	2,000	3.7
Huemules	12.08.71 01.73	43° 51' S 73° 06' W	250[8]	Lahars	35	ocean	large[9]	unknown	> 10,000	unknown	100	⩾ 10
Paine	01.82 12.82 03.83	50° 50' S 73° 08' W	190	Emptying of lake retained by glacier	32	lake	small	250	350	20 days	100	3.5

[1] Potentially large losses if the reservoir had been full. [2] Operation of hydropower plant interrupted. [3] 37 deaths. Destruction of hydropower plant; interrupted construction of other. [4] Actually eight floods in 19 days. [5] Total of four gullies on the flanks of the Villarrica vulcano. [6] 15 deaths. Destruction of houses and several bridges. [7] Small damage due to preventive measures. Partial flooding of city of 50,000. [8] Altitude of the front of the glacier. [9] Devastation of sparsely inhabited valley. Several deaths. [10] Assuming hydrograph is triangular. [11] Total value of four gullies. * Length of reach where non-meteorological flood is larger than 100-year flood. ** Description of site where non-meteorological flood becomes less than 100-year flood. V Estimated runoff volume. Q_{max} Estimated peak discharge at initial point of flood. Time, Basis of hydrograph at initial point of flood. Q_{100}, 100-year meteorological flood at initial point.

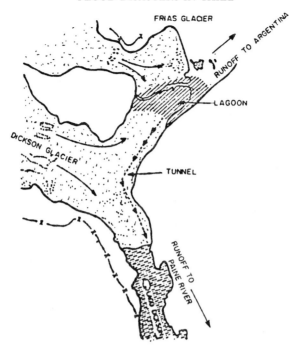

FIGURE 2 *Rio Paine flood*

According to the studies carried out for the Rio Paine flood (Peña and Escobar, 1983), in an area where two glaciers converge, the flow from one glacier to the other was obstructed and a lake of some 250×10^6 m^3 was impounded. Eventually this lake built up enough head to force through a tunnel under the Dickson glacier, reversing the direction of normal drainage (Figure 2).

It is worth noting that in this region the Patagonian ice cap has been withdrawing at a steady rate of an average 180 m per year between 1978 and 1983 with consequent changes in the overall balance of the system. The existing conditions favour large accumulations of water, as is reflected in the number of existing fringe lakes (Peña and Escobar, 1987b). Similar events elsewhere in Patagonia may have occurred unnoticed because there is hardly any permanent population in this region and the existing large lakes downstream of the glaciers are able to mask these comparatively small floods.

In the case of the Rio Cachapoal flood, the front of the glacier was covered with debris which allowed subglacial drainage of several small tributaries. At the time of the flood, the glacier had advanced to block the original drainage causing ponding at several places along the flanks (Rolando, 1981). These ponds broke out successively to give a 19 day sequence of floods each typically 1.5 to 2.0×10^6 m^3 in volume.

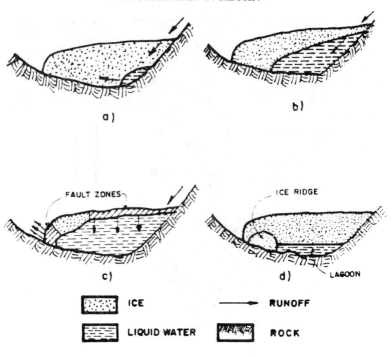

a)

b)

FAULT ZONES

c)

ICE RIDGE

d)

LAGOON

| ICE | | RUNOFF |
| LIQUID WATER | | ROCK |

FIGURE 3 *Rio Manflas flood*

(c) Formation of a subglacial lake. The flood of the Rio Manflas, in northern Chile (Table 1, case 1), which originated in the Rio Seco de los Tronquitos glacier at 5,200 m altitude, belongs to this category (Peña and Escobar, 1987a). Owing to the altitude and the extreme aridity of this region, glaciers can only exist in areas where the topography favours accumulation of wind-blown snow. These glaciers are small, very cold and show little activity.

The movement of the mass of ice is very slow. Under these conditions, the impounding of the volume of water observed (5×10^6 m^3) is quite surprising. Field studies led to the conclusion that a subglacial lake had formed in which water in the liquid phase was kept isolated from the external environment. The energy required to melt the ice and form the subglacial cave was supplied by the snowmelt water which warmed up the rock substratum. Figure 3 shows a scheme of the possible development of the event.

Two types of emptying process have been recognized:

(a) Gradual widening of an initial channel caused by heat transfer from the flowing liquid water to the ice, resulting in the gradual emptying of the reservoir. The physics of this process were studied by Nye (1976) who applied the equations of mass and energy conservation, of plastic flow of ice, of the hydraulics of the water duct and of the heat transfer that activates the phenomenon. It is thought that the floods caused by the Juncal (Rio Olivares), Cachapoal and Dickson glaciers (cases 2, 4 and 8) belong to this type. Of the eight floods considered in this paper, these three had the lowest peak flow.

In the case of the Dickson glacier flood, a model based on a simplified form of the Nye equations was able to explain the apparent large discrepancy between the volume of the flood (250×10^6 m^3) an the observed peak flow of only 350 m^3/s (Figure 5) (Peña and Escobar, 1983). Emptying of the Dickson glacier lake was initiated when the water reached about nine-tenths of the thickness of the ice, a value that is similar to those values recorded in other countries (Young, 1985).

(b) Structural failure of an ice wall. It is thought that the Rio Manflas flood from the Rio Seco de los Tronquitos glacier (case 1) belongs to this category (Figure 3), (Peña and Escobar, 1987). A similar case is documented for the rupture, on 12 July 1892, of the Tête-Rousse glacier in the French Alps (Lliboutry, 1964). The Rio Manflas flood was extremely large, with discharge at the foot of the glacier of 11,000 m^3/s. Peak discharge decreased rapidly to 30% of the initial discharge during the first 15 km of the channel, in particular in a reach where the slope is of only 3% (Figure 4). Further downstream the rounding of the peak continued at a slower rate and it was reduced to 10% of the initial value at 70 km. The flow velocity at the beginning is estimated at over 10 m/s, gradually reducing to values of the order of 5 m/s.

Reports from other countries in the Andes refer to floods caused by emptying of moraine lakes that had formed during the recent process of glacier withdrawal (Lliboutry et al., 1977). Similar cases have not been recorded in Chile.

Floods Generated by Volcanic Activity

Of the 15 volcanoes that erupted in Chile during the past 40 years, only three are known to have produced significant lahars: the Villarrica eruption (Gonzalez-Ferran, 1972; Marangunic, 1974), the Hudson eruption (Fuenzalida, 1974), and the Calbuco eruption (Klohn, 1963). In all these cases the generation of lahars was associated with the presence of glaciers and snow. Only the Villarrica and Hudson lahars, which produced the largest damage, are discussed in this paper.

The Villarrica volcano is very active, with eruptions in 1948, 1963/64, 1971 and 1984. The eruptions of 1948, 1963/64 and 1971 were associated with

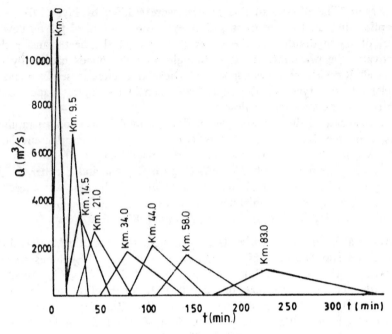

FIGURE 4 *Rio Manflas flood routing*

lahars that resulted in loss of life, the destruction of a village and other material losses. Only the better documented 1971 eruption is discussed here. The volcano (2,840 m) separates the Villarrica and Calafquen lakes, both at some 200 m altitude, and its lahars slide along gullies on the flanks of the volcano to end their course in one of the lakes. The large water mass of these absorbs the shock which is not propagated further downstream.

As described by Gonzalez-Ferran (1972) and Marangunic (1974), in February 1971 the ice on the flanks of the volcano was deeply fragmented and newly formed water pools indicated that the volcano was undergoing an active phase. On 29 October 1971 there were explosions in the central crater and on 29 November the first lava flow was recorded. Lava flow gradually increased until 20 December and the lava streams descended to the 2,000 m line. The flow of lava resulted in partial melting of the ice cover creating a 20 to 40 m-deep gully in the glacier. On 29 December a fracture of some 4 km appeared on the upper cone, ejecting a lava sheet which slid over the ice at considerable speed while fanning open until its front had descended some 400 m. Large quantities of water produced at the interface between lava and ice generated lahars that reached the lakes in less than one hour, following four different gullies. The maximum speed of the lahars was estimated to be about 10 to 20 m/s. The total volume of lava ejected during this eruption was estimated at 30×10^6 m^3. Not much pyroclastic material was produced.

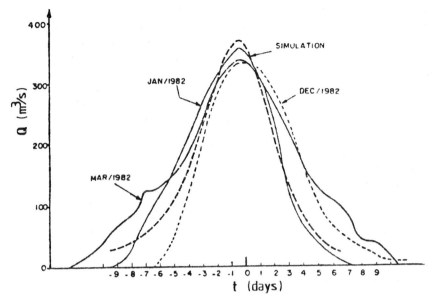

FIGURE 5 *Rio Paine flood hydrograph*

According to the water level records of both lakes, a volume of some 25×10^6 m^3 entered the lakes in about four hours. On this basis and assuming a triangular hydrograph, the maximum discharge of the lahars would have been of about 3,500 m^3/s.

The Hudson volcano (2,500 m), which has generated large lahars on the Rio Huemules, is in the Patagonian part of the Cordillera. The volcano is a roughly circular caldera of 9 km diameter which before the eruption was completely filled with ice. The ice mass fed two glaciers of which the larger projected to the Northeast into the Rio Huemules valley, and the smaller towards the Rio Ibañez in the Southeast. This volcano, situated in an uninhabited and inaccessible region, was discovered only in 1970 (Fuenzalida and Espinoza, 1973) and unexpectedly became active on 12 August 1971.

According to research by Fuenzalida (1974), the eruption of the Hudson volcano was very violent and expelled large quantities of volcanic ash which reached the Atlantic coast several hundred kilometres East. No lavas or coarse pyroclasts were observed. The eruption was initiated by an explosion followed by a large lahar that devastated 80% of the vegetation in the valley and killed most of the few inhabitants of the place. The lahar was composed of water, large ice blocks and a large quantity of volcanic material, in addition to logs and other debris. It sped down the Rio Huemules valley for 35 km, reaching the Pacific in the Elefantes Fjord. The day following the eruption the fjord was

251

covered with floating pumice for some 40 km. According to the information provided by Cevo (1978), the peak discharge of the lahar was larger than 10,000 m^3/s. It is thought that the lahar was generated by partial melting of the ice in the caldera added to the tensions introduced by the eruption resulting in the outburst of the accumulated water to the Rio Huemules valley. It is not known whether the ice was superficially melted by hot material expelled from the volcano. It is interesting to note that on photographs taken in 1945, there were fractures on the surface of the ice indicating abnormal thermal activity at its base. Also, in 1970 atypical behaviour of the Rio Huemules was observed, with sudden variations of discharge, possibly due to volcanic activity in the presence of ice.

In January 1973, without any new eruption of the Hudson volcano, a second catastrophic flood was recorded in the river, this possibly originated through an outburst of water accumulated in the caldera following melting of ice due to geothermal activity. This flood would be a typical case of Jokulhlaups, similar to some of those mentioned by Clague and Mathews (1973). As a consequence of the Hudson volcano activity, in February 1973, 80% of the ice that was originally in the caldera had disappeared (Fuenzalida, 1974).

Floods Generated by Breach of Water-Retaining Landslides Resulting from Seismic Activity

Earthquakes in mountainous areas are usually followed by landslides and falling rocks which tend to obstruct natural drainage with the ensuing risk of breach followed by sudden outflow of the accumulated water. During the past 40 years there have been some 15 large earthquakes in Chile (measuring over 7 on the Richter scale). Fortunately, owing to the high slopes of the narrow valleys, landslides have not been sufficiently large to retain large amounts of water, and only one exceptionally large flood that owes its origin to this phenomenon is known: the Rio San Pedro flood caused by the breach of a landslide triggered by the large earthquake of 22 May 1960 (8.7 on the Richter scale), which had retained the outflow of Lake Riñihue (case 6).

According to the information available (ENDESA, 1960), three large landslides obstructed the Lake Riñihue outlet. The largest of these, 4.5 km downstream of the lake outlet, had a volume of 34×10^6, was raised 43 m above the original river bed and filled 2 km along the length of the river. The earthquake that triggered the process unsettled the precarious stability of the canyon that the Rio San Pedro had excavated in sediments formed during the latest glaciation. The dam formed by the landslide caused the lake level to rise. The lake accumulated an additional volume of $2,000 \times 10^6$ m^3 in three months before overflowing and eroding the earth wall.

Confronted with the disaster potential of the situation, the authorities carried out works to divert the intitial overflow and built a bypass on stable

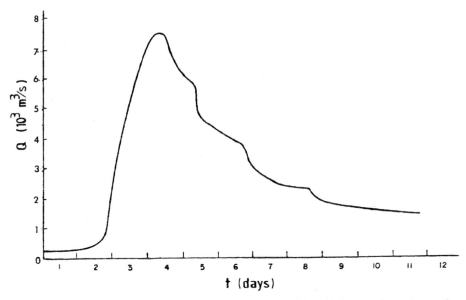

FIGURE 6 *Rio San Pedro flood hydrograph shows peaking 40 hours after release of main flood*

ground so as to avoid the water flowing over the loose erodible landslide material. Works were also carried out to heighten an existing rock control to raise the water level upstream. These measures succeeded in reducing the peak flow, originally estimated at 15,000 m^3/s, to the actually recorded 7,500 m^3/s.

The dam-breaking process was initiated slowly and under controlled conditions. A cascade was formed in a sector of more resistant ground, in such a way that downstream of this point there appeared a 15 m waterfall when the discharge was 300 m^3/s. When this control was disabled because of flow over softer ground, the main flood was unleashed and the peak reached within 40 hours (Figure 6). The flood resulted in partial submersion of the city of Valdivia (50,000 inhabitants), but losses were not large because enough warning had been available.

Floods Generated by a Landslide Followed by the Flow of Debris

Mud and debris flow has had an important role in shaping the landscape in a large part of Chile and constitutes most of the quaternary filling of many valleys, particularly in the arid and semi-arid parts of the country (Segerstrom, 1964; Golubev, 1969; Marangunic et al, 1979). Currently the frequency of these phenomena is high and has been correlated with the occurrence of periods of high rainfall and with particular geological formations (Hauser,

253

1985). However, most frequently the flow of debris moves only a short distance before it settles and therefore has only local importance. In the 29 November 1987 flood of the Rio Colorado, in the Andes near Santiago, sliding of a slope initiated a flow of debris which, through admixture of water on its way was converted into a large river flood with significant effects in a 59 km reach of the river bed. About 30 km downstream of the source of the event a construction camp was wiped out with a loss of some 30 lives. The event is still being investigated and it is expected to have more complete information in the future.

According to preliminary studies (Direccion General de Aguas, 1987; Chilectra, 1987), the event started in the headwaters of a minor tributary of the Rio Colorado, the Parraguirre brook, where there was a landslide of some 3.5×10^6 m^3. The landslide was initiated at about 4,700 m altitude and 1,200 m above the valley floor, so that a mass of soil and rock acquired large energy. This allowed it to jump a rock outcrop of over 50 m and turn 110° before joining the direction of the bed of the Parraguirre brook.

The immediate reason for the landslide is related to the exceptional snow accumulation in 1987, the second largest in the past 50 years, and to the high rate of snowmelt in the days preceding the landslide. The annual maximum discharge of the rivers in this region is normally caused by snowmelt at this time of the year.

The first 14 km of the path were in the narrow canyon of the Parraguirre brook with 7% slope, a discharge of perhaps over 10,000 m^3/s and speeds estimated at over 15 m/s. In this reach the debris flow would have incorporated large amounts of water available from the snow and from the partially saturated soil, and of mud resting on the slopes of the valley, so that the flow of debris became a hyper-concentrated water flow. There were large waves in the flow in accordance with the changes in the cross-section of the channel. After joining the Rio Colorado, which has a 2% to 3% slope in this section, the mean velocity was of the order of 8 to 10 m/s, with a peak discharge reduced to 1,000 m^3/s after a total course length of 59 km.

At the confluence of the Rio Colorado with the Rio Maipo, the changed hydraulic conditions resulted in a deposit of sediments with a thickness of some 5 m.

It is interesting to note that the flood had two peaks as can be clearly observed in the water level record 220 km downstream of the origin (Figure 7). According to witnesses, the second peak was much richer in sediments than the first. The reasons for this are not known.

General Analysis and Conclusions

The joint analysis of the non-meteorological floods discussed here suggests the following comments:

(1) In a country like Chile, with high mountains and glaciers, seasonal snow

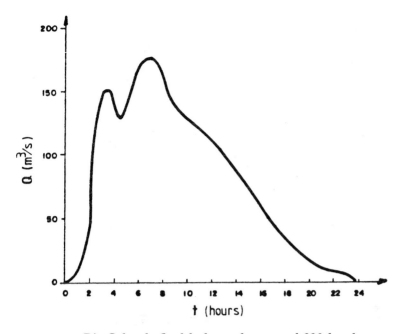

FIGURE 7   *Rio Colorado flood hydrograph measured 220 km downstream showing two peaks*

cover and high seismic and volcanic activity, the frequency of these phenomena is sufficiently high to be taken into account in development planning, particularly in relation to hydraulic structures that need a high degree of security and in relation to permanent human settlements. It is interesting to note that almost all non-meteorological floods in the years under consideration were generated on a basis of water stored in solid form (snow or ice) or in liquid form behind an ice dam, thus showing the importance of solid precipitation in the generation of these phenomena.

(2) The form of the hydrographs of these floods depends essentially on the physical processes at their origin and therefore the volume of the flood is not directly correlated with the peak discharge. This can be appreciated in the volume vs. peak discharge graph of Figure 8 when comparing floods generated through a continuous process of enlarging an initial duct or channel (cases 2, 4, 7 and 8 of Table 1), and floods related to a sudden violent phenomenon (Table 1, cases 1, 3, 5 and 7).

(3) Because of the diversity and complexity of the real-world situations, simple empirical methods are not sufficient to describe these floods. In the case of floods generated by the emptying of lakes retained by glaciers it can be seen in Figure 8 that the empirical relation of Clague and Mathews (1973) does not agree with the observed values. At the same time it can be

255

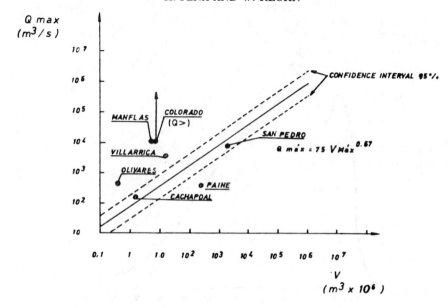

FIGURE 8 *Flood volume vs. peak discharge*

seen that mathematical modelling of the process of emptying ice dams carried out by Nye (1976), using simplified solutions (Clarke, 1982) or with numerical procedures (Peña, 1985) gives better results. The limitations inherent to the empirical methods can also be shown when applying to the Rio San Pedro a relation given to estimate the peak discharge related to break of an earth wall (Mc Donald and Langridge-Monopolis, 1984). In that case the observed peak discharge was four times less than predicted by the empirical relation.

(4) In all these events the sediment transport was much higher than normal, and in three cases (Rio Manflas, Rio Colorado and Volcan Villarrica) it is thought that sediments were a significant part of the flood volume.

(5) Generally, these floods caught the population unaware but they could have been partially forecasted if adequate programmes for monitoring glaciers and volcanoes had existed. However, the floods generated by emptying of a subglacier lake (Rio Manflas) and by debris flow as in the Rio Colorado, are very difficult to forecast and therefore particularly dangerous.

(6) The relevance of non-meteorological floods for hydraulic works depends both on the characteristics of the works of the flood itself. In relation to the latter aspect, it is important to know the type of event that may be expected, its frequency, volume, peak discharge at the site where it is generated, its routing parameters to the site of the hydraulic works and the possibilities for forecasting the event. To evaluate all the facets of the

problem it is important to achieve a better understanding of the physical processes involved and of their mathematical modelling, so as to have available powerful and flexible analytical tools able to tackle the diverse situations possible in nature.

Acknowledgement

The authors are indebted to Dr John B. Miller for his invaluable assistance in casting this paper into its final form.

REFERENCES

CEVO, J. (1978) Informe preliminar sobre la erupcion del volcan Hudson o Huemules, Trapananda No. 1, Coyhaique

CLAGUE, J. and MATHEWS, H.W. The magnitude of Jokulhlaups. *J Glaciol* 12: No. 66

CLARKE, G. (1982) Glacier outburst floods from Hazard Lake, Yukon Territory, and the problem of flood magnitude prediction. *J Glaciol* 28: No. 28

CHILECTRA (1954) Minuta interna acercas de la crecida del rio Olivares, Santiago

CHILECTRA (1987) Informe preliminar sobre el aluvion del 29 de noviembre de 1987 en el area del Proyecto Alfalfal, Santiago

DIRECCION GENERAL DE AGUAS (1987) Informe sobre el aluvion del 29 de noviembre de 1987 en el rio Colorado, Santiago

ENDESA (1960) Memoria Operacion Riñihue, Santiago

FUENZALIDA, R. (1974) El Volcan Hudson. Symposium Internacional de Volcanologica, Santiago

FUENZALIDA, R. and ESPINOZA, W. (1973) Hallazgo de una caldera volcanica en la provincia de Aysen. *Rev Geol Chile* No. 1

GOLUBEV, G. (1969) Avalanchas y corrientes de barro en Chile. *Informaciones Geograficas*, No. 17, Santiago

GONZALEZ-FERRAN, O. (1972) La reciente erupcion del volcan Villarrica. Primer Symposium Cartografico Nacional, Santiago

HAUSER, A. (1985) Flujos de barro en la zona preandina de la region metropolitana: caracteristicas, causas, efectos, reisgos y medidas preventivas. *Rev Geol Chile* No. 24

KLOHN, E. (1963) The February 1961 eruption of Calbuco Volcano. *Bull Seismol Soc Am* 53: No. 6

LLIBOUTRY, L. (1955) *Nieves y glaciares de Chile.* Editorial Universitas, Santiago

LLIBOUTRY, L. (1964) *Traité de glaciologie.* Masson, Paris

LLIBOUTRY, L., MORALES, B. and SCHNEIDER, B. (1977) Glaciological problems set by the control of dangerous lakes in Cordillera Blanca, Peru. *J Glaciol* **18:**

McDONALD, T. and LANGRIDGE-MONOPOLIS, J. (1984) Breaching characteristics of dam failures. *Proc ASCE J Hydraul Eng* **110:** 567–586

MARANGUINC, C. (1974) The Lahar provoked by the eruption of the Villarrica volcano in December 1971. Symposium Internacional de Volcanologica, Santiago

MARANGUNIC, C., MORENO, H. and VARELA, J. (1979) Observaciones sobre los depositos de relleno de la depresion longitudinal de Chile entre los rios Tinguiririca y Maule. Segund Congreso Geologico de Chile, Arica

NYE, J.F. (1976) Water flow in glaciers: Jokulhlaups, Tunnels and Veins. *J Glaciol* **17:**

PEÑA, H. and ESCOBAR, F. (1983) Analisis de una crecida por vaciamiento de una represa glacial. VI Congreso Nacional de la Sociedad Chilena de Ingenierai Hidraulica, Santiago

PEÑA, H. (1985) Simulacion de crecidas producidas por vaciamiento de represas de hielo. VII Congreso Nacional de la Sociedad Chilena de Ingenieria Hidraulica, Santiago

PEÑA, H. and ESCOBAR, F. (1987a) Analisis del aluvion de mayo de 1985 del rio Manflas, cuenca del rio Copiapo. VIII Congreso Nacional de la Socidad de Ingenieria Hidraulica, Santiago

PEÑA, H. and ESCOBAR, F. (1987b) Aspects of Glacial Hydrology in Patagonia: Glaciological Studies in Patagonia, 1985–1986. *Bull Glacier Res* (Data Center for Glacier Research, Japanese Society of Snow and Ice)

ROLANDO, G. (1981) Crecidas extraordinarias, Rio Cachapoal, Nota interna, Division El Teniente, CODELCO-Chile

SEGERSTROM, K.L (1964) Mass wastage in North-Central Chile. XXII International Geological Congress, India

YOUNG, G. (ED.) (1985) Techniques for prediction of runoff from glacierized areas. IAHS Publication No. 149

Hydrological Impacts of Earthquakes, Landslides and Avalanches

====

R.G. CHÁVEZ

Secretariat of Agriculture and Hydraulic Resources,
Mexico

Introduction

NATURAL phenomena have always aroused human curiosity, admiration and scientific interest. Some of them are also looked upon with fear, because the colossal amount of uncontrollable energy which they display sometimes causes anything from slight material damage to devastation, loss of many human lives and substantial, permanent environmental changes. Earthquakes, landslides and avalanches are among these potentially destructive phenomena which, by generating the movement of gigantic rock masses, have various impacts, some of them of a hydrological nature.

Over the last few decades, which have seen spectacular scientific and technological development, much progress has been made in our knowledge of the origin and controlling factors of these phenomena: the earth sciences have made in-depth studies of the natural processes associated with the planet's evolution which cause the vast majority of earth movements; increasingly complex and sensitive instruments have been designed to monitor them in detail; and a complex theoretical basis has been developed to describe them in mathematical terms.

This spectacular progress has enabled us to define the regions or sites which are most prone to these phenomena, and, in some cases, to predict their occurrence with a certain margin of error; however, despite this, the movements in question continue to have damaging, sometimes even catastrophic effects because, after ignoring warnings and precursory signs, people fail to take the relevant preventive action to attenuate their negative impact.

It has, moreover, been demonstrated that human activity also produces potentially destructive earthquakes, landslides and avalanches, although

259

presumably on a very small scale in comparison with those of natural origin, which has given rise to the conception of futuristic projects to control or attenuate the intensity of the natural ones.

The present paper contains a brief summary of current knowledge of such phenomena, a description of their main impacts with emphasis on those of a hydrological nature, and recommendations for various types of action to prevent human and material losses.

The Phenomena

A short survey of the causes and mechanisms of earthquakes, landslides and avalanches is necessary as a basis for considering their impacts and formulating recommendations to prevent or, at least, attenuate their harmful effects.

Earthquakes

An earthquake is a vibration and oscillation of the earth produced by a temporary alteration in the balance of rocks on, or below the surface. The most common natural causes are shifts in the earth's crust, volcanic activity and the collapse of rock masses. The vast majority of severe earthquakes are caused by the former, i.e. tectonic movement. It will be recalled that earthquakes are surface manifestations of seismic waves produced by friction, shock or subduction of the plates forming the earth's lithosphere.

Volcanic eruptions and the collapse of rock masses also produce vibrations which are perceptible sometimes at distances of up to several kilometres. However, in terms of energy, these seismic sources are insignificant by comparison with tectonic movements. In some cases, they can add the necessary momentary impulse to precipitate phenomena of another type (for example, landslides or avalanches), which were already imminent. In general, this also applies to artificial explosions, whose effect is relatively local.

On the other hand, variations in groundwater pressure may produce tremors of significant proportions, when they disturb the balance of rock masses. Indeed, the theory of regional flow systems postulates, and observations made in oil wells confirms that, at great depth, water is subjected to very high pressure; consequently, because of the mechanism described in Mohr-Coulomb's theory, the actual pressure and shearing resistance are relatively low, which explains the observed displacement of immense blocks of rock (several kilometres thick by hundreds of kilometres long) along fault planes with slight inclination (see Figures 1 and 2).

The above-mentioned mechanism also explains the origin of earthquakes recorded in areas where interstitial pressure is artificially increased. Cherry and Freeze (1979) quote two cases of seismic activity provoked by the injection of water into the subsoil through deep wells, to eliminate residual radioactive

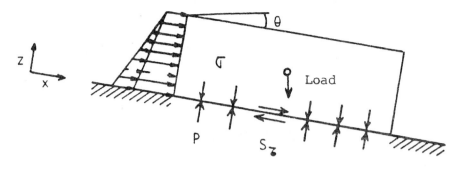

FIGURE 1 *The dominant role played by groundwater pressure in the equilibrium of blocks of rock*

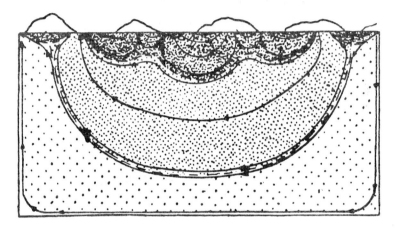

FIGURE 2 *Regional flow system (Toth, 1963) which includes several basins; at some depth the water is subjected to great pressure*

liquids and to extract oil deposits by the secondary recovery method. In both cases, it was observed that seismic movements began shortly after injection was started; close correlation was noted between the magnitude and frequency of earthquakes and fluctuations in hydraulic pressure; and it was demonstrated that the location of the epicentres coincided approximately with the injection site (see Figure 3).

Such experiments and the results of theoretical analysis have confirmed that groundwater pressure plays a very important role in fault movements, giving rise to the idea that catastrophic earthquakes can be prevented by artificially induced pressure changes, making it possible to liberate in a controlled manner the growing tectonic forces generated on the fault planes. Experiments have moreover been carried out showing the feasibility of this idea. The results

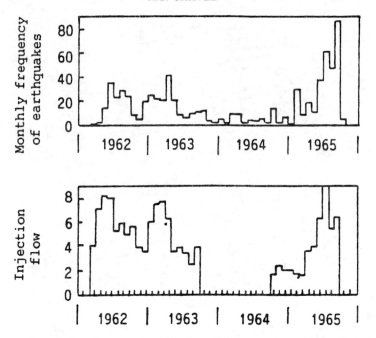

FIGURE 3 *Correlation between the injection of waste water and the frequency of earthquakes in Denver (according to Evans, 1966)*

obtained suggest that detailed, continuous observation of the behaviour of water levels in deep wells can provide indices which will help in earthquake forecasting.

Landslides

On sloping ground, the unconsolidated surface layer and consolidated underlying rock are subject to the action of gravity which, in certain conditions, can generate mass movements which are one of the geological processes modelling the planet's relief.

The most common causes of slides of rock masses are the force of gravity and groundwater. Gravity acts through the weight of the wedge of material which is liable to slide; any natural or artificial variation in this weight can affect the conditions of stability either favourably or adversely. An increase, by the addition of a load, favours the slope's stability when applied at the bottom, and reduces it when applied at the top. However, if the wedge loses weight, by erosion or artificial excavation, the reverse is the case.

Generally, groundwater is a determining factor for landslides. The moisture which occupies only part of the pores confers a certain cohesion on the unconsolidated material, through the surface tension of the film of water

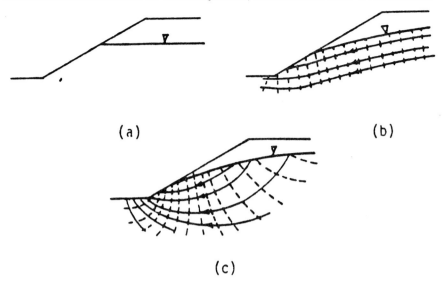

(a)

(b)

(c)

FIGURE 4 *Piezometric distribution commonly assumed in engineering works (a) and (b); distribution which is common in natural slopes situated in discharge areas (c)*

adhering to the grains; when the infiltrated water fills the pores, expelling the air, it breaks the surface tension and, consequently, reduces the material's cohesion while increasing its weight. The most important hydraulic factor in this process is pressure, because it directly influences the material's shearing resistance (Cherry and Freeze 1979).

Of all the factors controlling landslides, interstitial pressure is the least well known because there is usually insufficient instrumentation in the field, whether through underestimation of the importance of instruments, technical difficulties or economic restrictions. Because of this, the piezometric distribution is estimated using methods developed in the field of soil mechanics. However, when the slopes are very large, this approach may lead to dangerous underestimation of hydraulic pressure, since common hydrodynamic conditions are frequently presupposed in engineering works, and not the particular conditions of the regional flow systems. Special attention should be given to this hydrogeological aspect when the slope in question is located in a regional discharge zone (see Figure 4).

The flow regime also plays a determining role in the landslide mechanism. In some cases, it is reasonable to assume this as estabished for practical purposes; but in the majority of cases it is clearly transitory, especially when certain climatic and geohydrological factors are combined. For example, in sloping land, intense rain can cause a complex system of saturated-unsaturated flow and a rise in the water table of a magnitude and duration which depend on the rain's characteristics, the topography and various geohydrological

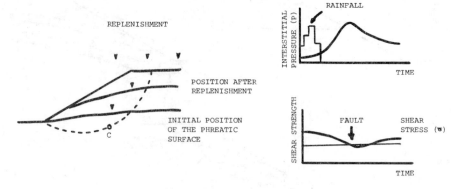

FIGURE 5 *The increase in interstitial pressure caused by infiltration can cause landslides*

factors (moisture content and hydrodynamic properties of the material). Because of the rise of the water table, the water pressure increases and the shearing resistance decreases proportionately throughout the plane of weakness and a landslide may then occur. This explains why many movements of this type take place after rainfall or snowmelt (see Figure 5).

The influence of water is still usually more difficult to analyze when slopes are formed by slightly fractured rock of complex structure because the distribution of stresses and water pressure is determined by specific structural breaks: faults, fractures and stratification planes, which makes it indispensable to make a detailed study of the structural geology and of the geohydrological conditions of the area in which the particular slope is located. Furthermore, because of the very low porosity of the slightly fractured rock, the infiltration causes big fluctuations in the water table, increasing the risk of landslides in the rock wedges during the rainy season (see Figure 6).

Besides gravity and water, which are the main causes of the vast majority of landslides, there are other contributing factors. As pointed out earlier, excavations at the base of a slope make the latter less stable, since, when the lateral support is removed, there is a reduction in the normal stress and shearing resistance which act on the potential landslide area. Strong vibrations caused by earthquakes, impacts or explosions can precipitate the movement of a slope which was already in a precarious state of equilibrium. Certain chemical changes in the material can also reduce shearing resistance.

Mohr-Coulomb's fault theory, conventional methods of slope stability analysis, and triaxial tests with their respective modifications and variants taking into account complex conditions constitute the theoretical and experimental tools to analyze most problems related to this type of movement (Juarez and Rico, 1980). Essentially, landslides of natural origin, whether slow

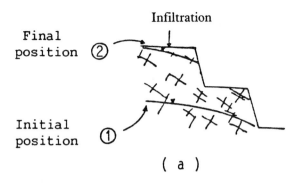

(a)

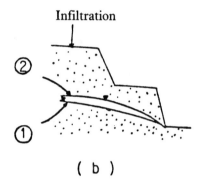

(b)

FIGURE 6 *Infiltration of the same volume of water causes much greater rises in the water table in (a) slightly fractured rock than in (b) porous material*

or fast, small or large, have the same cause, and are analyzed with the same tools as those in engineering works (see Figure 7).

Both in the unconsolidated detrital cover and in the very fractured and weathered rocks, which form a hydrologically continuous medium, the slippage plane resembles a cylinder since the detached block has a circular outward motion along what is known as a 'slope fault'. Frequently, a depression forms in the upper part of the fallen block which collects rainwater and facilitates its infiltration, thus helping the movement to continue (see Figure 8). The material's anisotropy with respect to its shearing resistance modifies the form of the landslide area, deflecting it from the circular arc. In stratified, rocky land, it is more common for blocks to become detached by sliding along a stratification plane, i.e. with a translation motion, whilst in slightly fractured rocks this area takes on very irregular forms determined by the fracturing pattern.

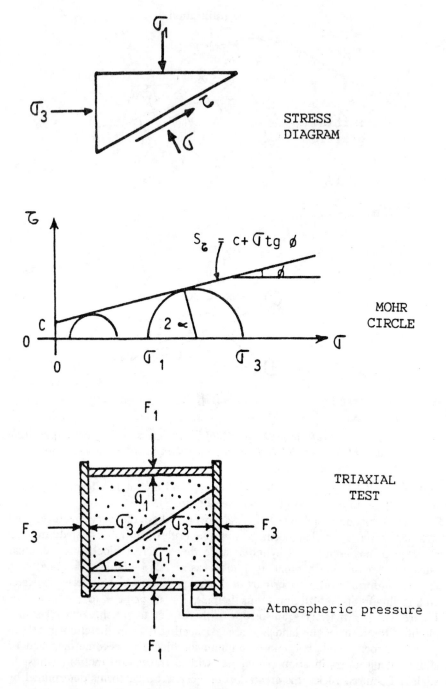

FIGURE 7 *Theoretical and experimental bases for the analysis of slope stability*

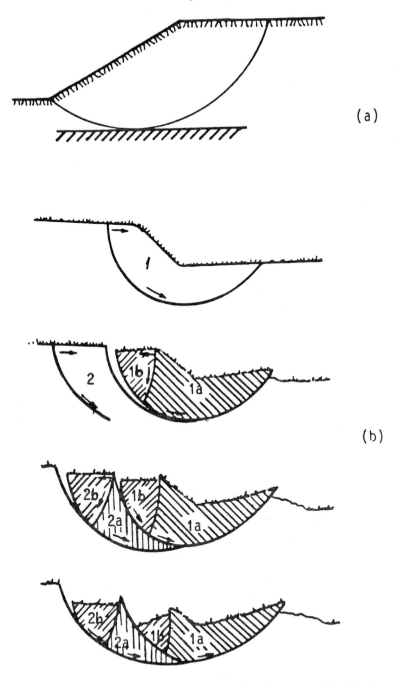

FIGURE 8 *(a) Rotative landslide of a slope; (b) fragmentation of the block removed by successive landslides*

Landslides can be fast or slow. In the former case, when they are spectacular and easily perceptible, they are less destructive. Their magnitude can vary from a local phenomenon on a single slope to the removal of the side of a mountain. Within this range, the most catastrophic are sudden slips of a rock mass along a plane of weakness. The technical literature contains various spectacular cases of movements of this type, most of which take place in land made up of marine sedimentary rock.

When the unstable material is detached by a flush of water generated by a violent storm or a thaw, a mudslide occurs, composed of rocks, soil and water, which progresses down the slope along channels and valleys forming a tongue of mud which can form balls of great size and carry them along. These mudslides transport gigantic volumes of material in alpine and desert areas with sparse vegetation and fast surface runoff.

The slow landslides are less perceptible and destructive than the fast ones, but, since they happen almost continuously over long periods of time, they transport much greater volumes of material than the fast, spectacular ones. In zones with tropical and temperate climates, the soil and underlying weathered layer form a plastic mass which flows slowly downhill, even on moderately sloping land protected by natural vegetation. The movement is facilitated by a high moisture content in the subsoil and other processes which detach the material and facilitate the work of gravity.

The so-called 'soil-flow' which is common in the upper latitudes, consists in the slow movement of the detrital cover which is alternatively frozen and thawed; like the landslide, it progresses downhill and water cannot percolate through the strata which are still impermeable because of the ice, and the surface layers become saturated acquiring the consistency of mud, which flows even on moderately sloping land.

Avalanches

In cold-climate mountain regions, avalanches bring gigantic volumes of snow to rivers and glaciers in addition to a very considerable solid load, made up of blocks of rock, detritus and trees which are picked up in the downward movement. The sliding is caused by gravity assisted by atmospheric factors and by natural or artificial vibrations from earthquakes, explosions and intense noise. Avalanches play an important role in the phenomenon of glaciation, since in a very short time they can add to a glacier volumes of snow which are equivalent to those precipitated over several years (Martinec, 1989).

Hydrological Impacts of the Phenomena

The phenomena covered in this paper have many impacts with very varied nature and intensity, from minor local damage to regional devastation with the loss of numerous human lives; some impacts are of a hydrological nature.

Earthquakes

Apart from the serious damage in urban areas, which is only too well known, a severe earthquake can have various indirect hydrological impacts which, in turn, cause secondary disasters. The most common hydrological sequels are: modification of the piezometric distribution or of the natural discharge conditions of aquifers, caused by the alteration of the underground geological structure which is sometimes reflected in sudden changes in water levels in wells and the discharge from springs; destruction of sandy strata, which also causes fluctuations in groundwater levels; pollution of aquifers which is favoured by the cracking of the land, etc. An earthquake often also has hydrological effects associated with landslides and cracks in dams or riverbanks (Yoshino, 1989).

Mexico is often subject to these phenomena: it is estimated that approximately 3% of the planet's seismic energy is liberated in this country; on average, an earthquake with an intensity of more than 7 on the Richter scale occurs once every two years, and, during the present century, eight earthquakes measuring 8 or more on the Richter scale have been recorded. This considerable activity is caused by the interaction of four tectonic plates: North America, Pacific, Cocos and Caribbean; the first two of these are displaced with different relative movements, giving rise to the seismic activity which affects the north-western part of Mexico. However, it is the convergence of the plates which is the main cause of the telluric movement; in particular the subduction of the Cocos plate below that of North America in the southern part of Mexico has caused many earthquakes with an intensity greater than 7.

Unquestionably, the most destructive of these occurred on 19 September 1985, affecting the most densely populated part of the country, Mexico City, a megalopolis of 18 million inhabitants, with extreme intensity. This earthquake was, in fact, made up of two movements of 8.1 magnitude, occurring 27 seconds apart, whose epicentres were located in the Pacific Ocean, some 800 km from Mexico City; on the following day, 36 hours later, another quake occurred which was 7.5 on the Richter scale. Among the main consequences were: more than 10,000 dead and a much greater number wounded, 412 buildings demolished and many more damaged, seriously damaged communication systems, a medical centre practically demolished, pollution of many aquifers by leaks into the sewerage system, power and water supplies cut off in large urban areas, etc., all of these effects being concentrated in the valley which Mexico City occupies, where the movement's intensity was accentuated by the subsoil's characteristics.

Within this valley, the only hydrological consequence of any importance was the pollution of the aquifers in some areas, brought about by the cracking of the earth and breakage of the sewerage pipes and canals for removing waste water; several landslides also occurred in the surrounding mountains, but without significant hydrological implications. On the other hand, in basins in

269

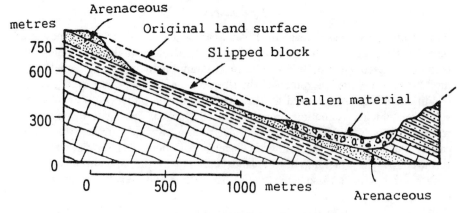

FIGURE 9 *Landslide in Wyoming*

the neighbouring state of Morelos, a sudden drop in water levels in wells scattered over a large area was observed immediately after the earthquakes, as well as the gradual reduction of the discharge from large springs, from a total discharge of more than 1,000 l/s to about 70 l/s.

Landslides

Landslides have various hydrological impacts, some of them being potentially destructive. Frequently, the material which is removed forms a dyke which obstructs the course of a surface stream; the water accumulates until it lowers or destroys the dyke, resulting in flooding (Peña and Klohn, 1989).

The technical literature contains a case which occurred in Wyoming in 1925. A block of 40 million m^3 fell off the side of a mountain because the percolated water reduced the cohesion between the strata; the rock mass slid along a slippage plane, crossed the valley, and smashed against the opposite side with such force that it produced a 'wave' 100 m high and 'backflow' forming a detrital layer 70 m thick; the water retained there gave rise to a lake 8 km long, which overflowed causing flooding downstream (see Figure 9, Leet and Judson, 1954).

The hydrological consequences are usually more serious when a large rock mass falls into a body of water, since the wave caused by the shock can produce destructive waves. A less harmful consequence is the change in runoff regime of a stream, which is caused by the large volume of sediment deposited in its channel.

Avalanches

The hydrological consequences of avalanches are essentially similar to those of landslides described above. Depending on factors which have been fully described by J. Martínez, the seasonal runoff regime is speeded up or slowed

Regolite

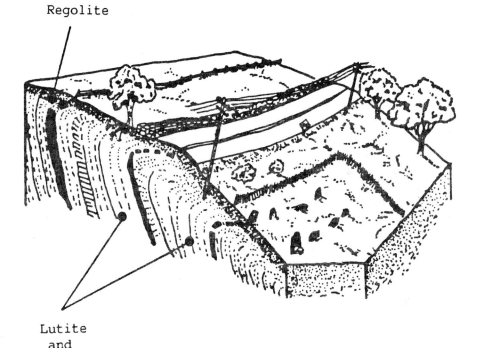

Lutite
and
arenaceous

FIGURE 10 *Landslide of the cover of regolite and deformation of the underlying strata*

down, in accordance with the thawing conditions of the snow added to a glacier; this modification can be favourable when it has a regulating effect (Martinec, 1989).

On the contrary, if the load of the avalanche obstructs the course of a river, the hydrological consequences are generally harmful and can be even catastrophic: the water is temporarily stored and later suddenly liberated, eroding the bank and giving rise to extraordinary flood waves with disastrous secondary effects. Similar impacts are produced by snow and ice masses falling into water bodies.

Recommendations to Attenuate the Harmful Hydrological Consequences of Disasters

Some of the hydrological consequences of earthquakes, landslides and avalanches in certain regions threaten human life and activities. The prevention of such consequences represents a formidable scientific and technological challenge requiring the co-ordinated participation of multidisciplinary groups.

271

It is generally supposed that the occurrence of these phenomena is unpredictable and therefore takes everyone by surprise; however, in reality, nearly all have precursory signs. For example, some earthquakes are preceded by abnormal fluctuations in groundwater levels or by gaseous emanations; the probability of a landslide is indicated by cracks growing in the ground and on engineering works, displacement of the soil profile, and inclination of tree trunks (see figure 10). Presumably, some animals perceive signs indicating the imminence of a natural phenomenon of some magnitude. Unfortunately, such indices are not observed in time or they are observed by witnesses who cannot interpret them, as happens with most of the rural population.

On the other hand, it is certainly true that, because of their extreme complexity, these phenomena are difficult to predict precisely; however, studies made in various scientific disciplines, and the statistics collected on this subject in various countries have enabled us to identify the regions which are most subject to them, as well as to draw up safety standards, design criteria and relevant emergency programmes to prevent or attenuate their consequences. With this objective in mind, the following specific recommendations are put forward:

- to map the zones which are most subject to these phenomena, with regard to their climatic, topographical, geological, hydrological and geohydrological characteristics. An interdisciplinary effort is indispensible for analyzing these aspects; they are often rightly studied with an emphasis on one or several specialized aspects, but the combination of their results is missing;
- to give special attention to the geohydrological aspects in the analysis of slope stability, since it has been demonstrated that groundwater plays a dominant role. In particular, careful consideration should be given to the structural geology and natural piezometric distribution in ground made up of stratified and fractured rocks;
- to develop a specific methodology to discover the underground hydrology of massifs;
- to install instruments in zones and engineering works which are most vulnerable to the phenomena in question, and set up continuous observing programmes in order to collect data to identify the beginning of one of them;
- to include in dam projects the simulation of a rise in the regional water table caused by infiltration in the reservoir and its influence on the stability of adjacent sloping terrain;
- to review the criteria used for the siting and design of engineering works and adapt them to the particular conditions of each zone;
- to make a study of the underground geological structure in earthquake zones in order to know the distribution of geological formations which are most affected by the passage of the waves;
- to promote the exchange of information, experience and new methodologies between specialists in the various disciplines related to the hydrology of disasters.

REFERENCES

CHERRY, J.A. and FREEZE, R.A. (1979) *Groundwater.* Prentice-Hall, New York

JUAREZ, E.B. and RICO, A.R. (1980) *Soil Mechanics* Vol III. Limusa, Mexico

LEET, L.D. and JUDSON, S. (1954) *Physical Geology.* Prentice-Hall, New York

MARTINEC, J. (1989) Hydrological consequences of snow avalanches. In: *Hydrology of Disasters* Starosolszky and Melder (Eds) James and James, London, pp. 284–293

PEÑA, H. and KLOHN, W. (1989) Non-meterological flood disasters in Chile. In: *Hydrology of Disasters* Starosolszky and Melder (Eds) James and James, London, pp. 243–258

YOSHINO, F. (1989) Hydrological consequences of earthquakes experienced in Japan. In: *Hydrology of Disasters* Starosolszky and Melder (Eds) James and James, London, pp. 274–283

Hydrological Consequences of Earthquakes Experienced in Japan

==

FUMIO YOSHINO

Public Works Research Institute, Ministry of Construction, Japan

Introduction

MAJOR changes of morphology caused by the phenomena of earthquakes, eruption of volcanoes, and slope failure by severe storms, cause abnormal movements of sediments and water. They affect hydrological phenomena in the field concerned.

Earthquakes or landslides directly cause disasters in the form of morphological changes, but their hydrological consequences are not directly recognized in the first stage of a disaster. After the earthquake or landslide, the soil mass supplied by slope failures blockades river valleys and produces dammed reservoirs. Dammed reservoirs are liable to collapse, which may cause secondary disasters, such as downstream flooding by the propagation of surge waves. Therefore, hydrological conditions are widely changed in river sedimentations. The effects cause the morphological change of river courses and therefore hydrological conditions of rivers and nearby groundwater bodies. Morphological changes have been observed in many earthquakes but their hydrological consequences were not clearly recognized in the past.

On the other hand, as landslides are triggered by a rise of groundwater level and a rise of pore pressure, hydrological phenomena have a direct effect on the occurrence of landslides. However, as the report concentrates mainly on the hydrological consequences of an earthquake or landslide, the direct effects of hydrological phenomena are not treated here.

The main hydrological consequences are recognized in morphological changes caused by fault movements, earthquakes or landslides. They include a change of drainage area of ground and surface water, an increase of sediment discharge caused by the increase of sediment supply from the collapsed area, or an increase in disaster by the collapse of slopes loosened by earthquakes at

TABLE 1

Hydrological aspects of earthquakes

Direct effects	Secondary effects	Effects on hydrological phenomena
Occurrence of faults	Discontinuity of soil layers	Change of drainage area Effect of water level on ground water
Slope failures	Blockage of river valleys	Propagation of surge in the valley and inundation Failure of dammed reservoir and flooding
	Sedimentation	Occurrence of debris flow Rise of river bed
Destruction of buildings	Failure of river banks	Flooding
	Dam failure	Occurrence of surge
Loosened slopes or cracks	Instability of slopes	Secondary slope failure
Soil liquefaction	Abnormality of groundwater condition	Groundwater level change
Tsunami	Flooding in estuaries	

the time of heavy storms, etc. As they cannot be expressed analytically, the hydrological consequences of earthquakes or landslides are evaluated through examples of these disasters in this report.

Hydrological Consequences of Earthquakes

Earthquakes produce large-scale faults, the destruction of buildings, large-scale slope failures, or loosened slopes and cracks which may not be recognizable. These effects are classified in Table 1.

Major hydrological consequences are caused by the blockade of river valleys, the occurrence of dammed reservoirs, and the collapse through overtopping of stored water, which may cause secondary disasters. These secondary disasters are classified in Figure 1 (Japanese Association 1986).

Table 2 shows examples of major slope failures caused by earthquakes experienced in Japan (Japanese Association 1986). Therefore, hydrological consequences of earthquakes are mainly recognized in secondary disasters by the blockade of river valleys and sedimentation problems.

275

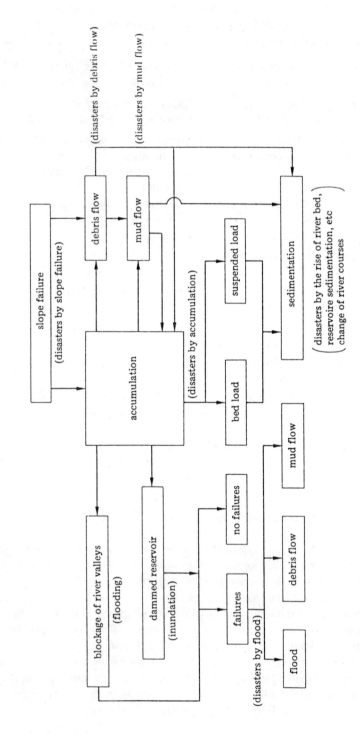

FIGURE 1 *Type of disasters with slope failure caused by earthquakes*

TABLE 2

Huge slope failures caused by earthquakes in Japan

Name of slope failure	Related river	Volume of sand mass (m^3)	Name of earthquakes	Magnitude	Date of occurrence
Oosawa	Huzi	7.5×10^7			19.08.1331
Kiunzan	Shou	2.5×10^7		7.9	18.01.1586
Ooya	Abe	1.2×10^8	Keichou	7.9	03.02.1605
Natate		4.0×10^7	Takata	6.6	20.05.1751
Kokuzouyama	Sai	$> 10^7$	Zenkouji	7.4	08.05.1847
Ootonbiyama	Zyouganji	4.1×10^8	Hietu	6.9	09.04.1858
Ontake	Kiso	3.6×10^7	Naganokenseibu	6.8	14.09.1984

Examples Experienced in Japan

The Zenkouzi earthquake

The Zenkouzi earthquake is famous for the blockade of the Sai river and the secondary disasters caused by the collapse of the dammed reservoirs in the river valleys. The earthquake occurred in 1847 and disastrous effects of the earthquake were counted in the death of more than 10,000 people and the loss of 20–30,000 households. More than 40,000 slope failures were counted. The largest occurred on the slope of the Kokuzou mountain and in the huge volume of sediments accumulated in the Sai river. The accumulated volume is estimated at about 700 m along its river course with a height of about 100 m. The huge dammed reservoir that occurred had a length of more than 40 km. Almost all the river flow was dammed for two weeks after the earthquake and about 30 villages along the Sai river were flooded. Three weeks after the earthquake, the dammed reservoir was overtopped by the flowing water caused by severe storms succeeding the earthquake. The stored water flowed into the Sai river and surge floods killed more than 100 people.

The Naganoken seibu earthquake

A large earthquake of the magnitude 6.8, with its centre on the south slope of Mt. Ontake, occurred on 14 September 1984. The earthquake created many slope failures in the Otaki river basin. The total amount of sediments were estimated at about 36 million m^3. The sediments moved down the tributary valleys in the form of debris flow. The survey established that the height of debris flow reached more than 100 m above the original river beds along the Denzyou river, one tributary of the Otaki river. The sediments accumulated in the Nigori river, which has relatively mild river bed slope. The length of the

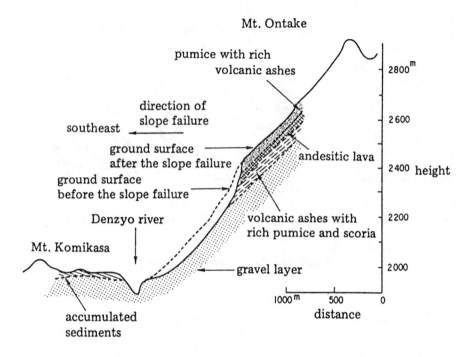

FIGURE 2 *Sectional figure of the slope failure on Mt. Ontake*

accumulated area of sediments was estimated at about 4.1 km and maximum width of the area was estimated at about 450 m. Their depth reached 20–50 m above original river beds.

Figure 2 shows the cross-sectional view of slope failure on Mt. Ontake. The sediments were widely accumulated along the river courses. Therefore, it was thought that the accumulated river bed was not easily moved and secondary disasters by the collapse of dammed reservoirs might not be expected.

The Japanese Government dispatched the survey team which directed necessary countermeasures in order to mitigate the secondary disasters by the earthquake and the necessary preventive works were undertaken in the field. The survey team concluded that secondary disasters were not expected from the overtopping of dammed reservoirs, nevertheless severe storms might occur and that work for discharging of waters from dammed reservoirs and preventive work against movement of accumulated sediment should be undertaken quickly.

Therefore appropriate measures were taken to prevent a disaster, and rehabilitation works along the Otaki river were commenced.

Figure 3 shows the results on sedimentation in the Makio reservoir, which is located about 10 km downstream from the confluence of the Nigori river on

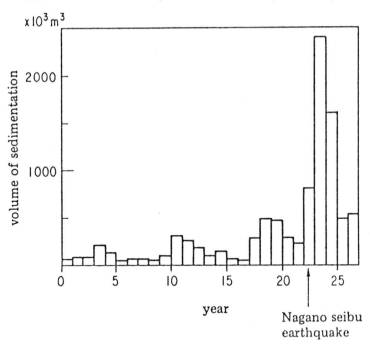

FIGURE 3 *Change of sediment discharge by the earthquake*

the Otaki river. The hydrological effect of the earthquake is clearly shown in the sediment discharges before and after the earthquake.

The Matsushiro earthquake (National Committee, 1969)
Outline of earthquakes

Frequent earthquakes occurred at Matsushiro town in Nagano prefecture from 3 August 1965 until the end of March 1968. The total number of earthquakes reached about 680,000 including about 61,000 earthquakes recognized by man. The largest magnitude was 5.2, which occurred on 3 February 1967. The depth of the earthquake centre was estimated at 10 km under the ground surface. These frequent earthquakes did not create large amounts of damage, but secondary effects of the earthquakes occurred near the town of Matsushiro. The worst damage was caused by the landslides which occurred in the suburbs of Matsushiro.

Abnormality within ground water

The earthquakes created many cracks in the surface of the ground. Figure 4 shows a general description of surface abnormalities by the earthquakes. Many springs occurred in the plain, but groundwater discharge decreased in many wells in the mountainous area eastward of the Tikuma river.

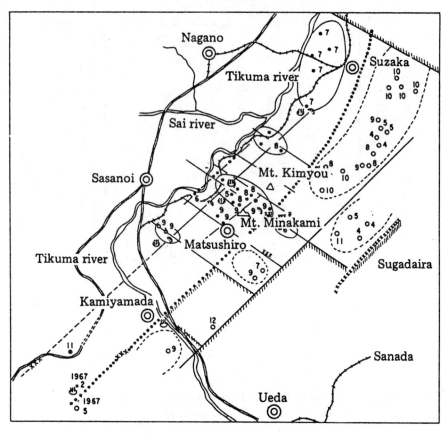

	remarks	
⌒	land slides	
ɪɪɪ	cracks	
•	springs (number shows the month of occurence)	
○	dry springs (number shows the month of occurence)	
T	craters	
♨	hot springs	
/////	no change of groundwater	

FIGURE 4 *Occurrence of groundwater abnormalities in Matsushiro region (by Mr Noboru Yamagishi, National Committee, 1969)*

Groundwater abnormality was recognized from May of 1966. At the village of Tanaka, water welled up from many places on 20 May 1966 and the area extended from north to south of the village. Until September, welling up could be seen in the villages near Tanaka, discharging a significant amount of water. The water was normal groundwater at first, but the amounts of CO_2 and Cl in the water gradually increased. Finally, water from hot springs emerged in some places.

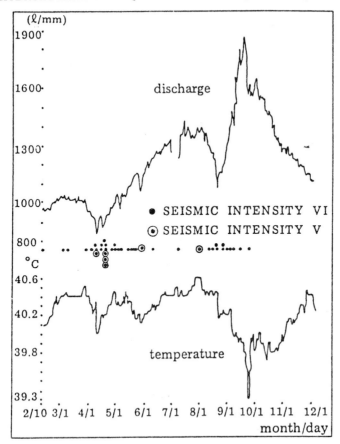

FIGURE 5 *Discharge and temperature of Kagai hot spring*
(by Mr Isao Kasuga, National Committee, 1969)

On the other hand, natural discharge from wells at Takimoto village gradually decreased from August 1966 and the water of Kiyotaki, which had had stable water discharge for many years (on the east slope of Mt. Kimyou), finally dried up. The same phenomena were observed eastward of Makiuti village.

Moreover, the discharge from hot springs increased at Kagai. Figure 5 shows the change of discharge and its temperature at Kagai hot spring. At the end of August 1966, the discharge from hot springs rapidly increased. The period coincided with the period of rapid increase of discharge from newly welled up places and also the period of large movement of ground surface. At the same time landslides occurred successively in the district of Makiuti.

The landslides directly related with the discharging of groundwater, which resulted in the secondary disasters caused by earthquakes. Discharging of

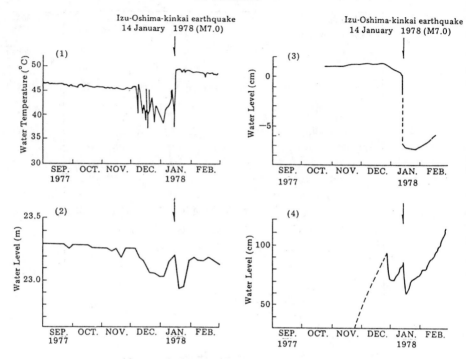

FIGURE 6 *Abnormal change of groundwater before and after the Izu-Oshima-kinkai earthquake (by Dr Masakazu Otake, Japanese Association 1986) (1),(2) Nakaizu, (3) Hunahara, (4) Omaezaki*

groundwater occurred in the area of cracked baserock. Therefore, it was concluded that the landslide was triggered by the groundwater discharged from confined aquifer.

Change of groundwater conditions by the Izu Ooshima Kinkai earthquake (Japanese Association, 1986)

The Izu Ooshima Kinkai earthquake occurred on 14 January 1978. A magnitude of 7.0 was observed. Before the earthquake, groundwater conditions of water level and water quality were observed and the changes are shown in Figure 6.

Observations revealed that the radon content of groundwater gradually decreased at Nakaizu from about three months before the earthquake and its abrupt change occurred after the earthquake. The same tendency was recognized in the temperature of hot springs at Nakaizu and groundwater levels at Nakaizu, Hunahara and Omaezaki, shown in Figure 6. The hydrological consequences of the earthquake are clearly recognized in the change of groundwater level.

Concluding Remarks

Hydrological consequences of earthquakes are directly recognized in secondary disasters such as failures of dammed reservoirs and changes of groundwater conditions. The main features of hydrological effects by earthquakes are classified in this report and their aspects are reported through examples experienced in Japan. However, the effects of earthquakes on hydrological conditions in drainage areas have never been studied extensively. The geomorphological changes will affect the movement of water in the drainage area and hydrological consequences may be recognized in the long term. These problems remain to be studied in the future.

REFERENCES

JAPANESE ASSOCIATION FOR DISASTER PREVENTION (1986) The forecast and countermeasures against secondary disasters (in Japanese)
NATIONAL COMMITTEE FOR COUNTERMEASURES AGAINST LAND SLIDES (1969) Matushiro earthquake and the land slide (in Japanese)

Hydrological Consequences of Snow Avalanches

—

J. MARTINEC

Federal Institute for Snow and Avalanche Research,
Weissfluhjoch/Davos, Switzerland

Introduction

EVERY YEAR, on the slopes of mountain basins, large volumes of snow are moved downwards by avalanches. The runoff pattern may be affected in various ways: Due to the changed melting conditions, the seasonal contribution to runoff is either accelerated or retarded. Depending on the local runoff regime, both eventualities can change the runoff either favourably or unfavourably. Another effect may occur if the snow is deposited across a water course, so that the flow is temporarily detained. The river damming can be aggravated if an avalanche contains rock and other solid material.

Computed Effect of Avalanches on Snowmelt and Runoff

The change of snowmelt conditions at the avalanche deposit as compared with the conditions in the zone of origin can be evaluated by a formula proposed by de Quervain (Martinec and de Quervain, 1975):

$$M = a_T \times T + M_R(1 - r) - G \qquad (1)$$

where M is the daily snowmelt depth in cm

a_T is a coefficient [cm °C^{-1} d^{-1}] not to be confused with the overall degree-day factor

T is the number of degree-days [°C × d]

M_R is the global radiation converted to the daily meltwater depth [cm]

r is the albedo as a decimal number

G is the net outgoing longwave radiation converted to the daily meltwater depth [cm]

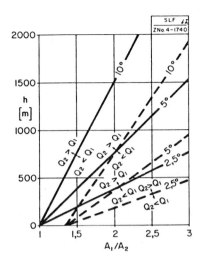

FIGURE 1 *Balance between the effect of altitude difference and of area on runoff for* $\gamma = 0.65°C\ 100\ m^{-1}$. *Dashed lines:* $a_1 = 0.45$, $a_2 = 0.65\ cm\ °C^{-1}\ d^{-1}$

Due to the temperature lapse rate, the number of degree-days is higher at the avalanche deposit than in the origin zone. Also, in most cases, the global radiation becomes more efficient due to a dirty snow surface and a reduced albedo. Consequently, the daily snowmelt depths in the zone of deposition are greater than in the zone of origin. On the other hand, the area of the avalanche deposit is frequently smaller than the area in the zone of origin affected by the avalanche. Thus the area of the snow cover surface exposed to melting is reduced.

Figure 1 illustrates the balance of these two contradictory effects. The snowmelt components of Equation (1) are summarized in an overall degree-day ratio assuming a value of 0.45 cm $°C^{-1}\ d^{-1}$. In the second example (dashed lines), the effect of a dirty surface of the avalanche deposit was taken into account by increasing this value to 0.6 cm $°C^{-1}\ d^{-1}$. The lines signify equal effects of the altitude difference and area for various air temperatures in the zone of origin as indicated in the figure. Equal meltwater volumes are expressed by the following equation:

$$a_1 \times T \times A_1 = a_2(T + \gamma \times h) A_2 \tag{2}$$

where a_1 is the degree-day factor for the snow in the zone of origin [cm $°C^{-1}d^{-1}$]
a_2 is the degree-day factor for the snow in the avalanche deposit
T is the number of degree-days [$°C \times d$]
A_1 is the area of the zone of origin
A_2 is the area of the avalanche deposit

285

    h is the altitude difference between the zone of origin and the zone of deposition [m]

    γ is the temperature lapse rate [°C m^{-1}]

If $a_1 = a_2$, Equation (2) is reduced to

$$T \times A_1 = (T + \gamma \times h)A_2 \qquad (3)$$

If the effect of the altitude difference h prevails, the snow displacement by an avalanche increases the snowmelt volume per day so that the daily contribution to runoff Q_2 is greater than the daily contribution to runoff Q_1 which would be available in 'no avalanche' conditions. If the area of the avalanche deposit A_2 is much smaller, due to terrain morphology, than the zone or origin A_1, this effect may be reversed: the snowmelt in terms of daily volumes is slowed down and major contributions to runoff from the avalanche-affected area are shifted to later stages of the snowmelt season. If A_2 is larger than A_1, the area effect as well as the altitude difference effect join in accelerating the snowmelt and runoff.

Evaluation of the Avalanche Effect on Runoff from Measurements

An artificially released avalanche on the slope of Brämabuel near Davos, Switzerland, offered a possibility to examine the hydrological consequences by measuring the redistribution of snow and the progress of the snow ablation in the changed conditions. The measurements had to be complemented by computations and estimates. The avalanche was triggered off on 25 February 1970 after a month of heavy snowfalls and brought down about 400,000 m^3 of snow (Figure 2).

The affected area and the longitudinal profile are shown in Figure 3. Figure 4 shows the distribution of snow before and after the avalanche in terms of the water equivalent in centimetres and, taking into account the areas of the respective elevation zones, in terms of the equivalent water volumes. Detailed evaluations of the snow cover on this date and in the following months have been published elsewhere (Martinec, 1971; Martinec and de Quervain, 1975). The snow ablation in the whole avalanche-affected area which measured about 480,000 m^2 was computed zone by zone and verified by periodic measurements of the remaining snow cover. Snowfalls between 25 February 1970 and the final disappearance of snow had to be also taken into account. The cumulative ablation from the avalanche-affected area (Figure 3) approached 400,000 m^3 water. This amount refers to all snow in this area while the 400,000 m^3 of snow dislocated by the avalanche corresponds to estimated 108,000 m^3 of water. As shown in Figure 5, the ablation in the avalanche-influenced area was ahead of that in the undisturbed snow cover nearby in April and May. In June and July, the undisturbed snow cover was

FIGURE 2 *The Brämabuel avalanche near Davos, 25 February 1970*
(photo L. Gensetter, Davos)

catching up. The final disappearance of snow in mid-July was almost simultaneous, although the last small remainders of the avalanche deposit outlasted the undisturbed snow cover by about two weeks.

Since the evaporation losses are small, the ablation amounts can be approximately considered as contributions to runoff from the area in question. Figure 6 shows these contributions totalized for intervals of 2–4 weeks, again in comparison with runoff contributions from an undisturbed snow cover. In April and May, the runoff contributions from the avalanche-affected area were higher than contributions from the undisturbed snow cover while in June and July they were lower. Whether the seasonal runoff pattern was thus influenced favourably or not can be evaluated only when this effect is extrapolated to a basin with measured runoff.

Extrapolation of the Single Avalanche Effect to a Mountain Basin

The avalanche occurred just outside the Dischma basin which is situated at 1,668–3,146 m a.s.l. and has an area of 43.3 km². According to observations and estimates, about 10% of this area was affected by avalanches similar to that examined. Since the examined area was 0.48 km², the total effect of these

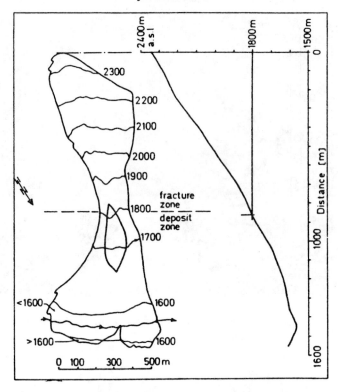

FIGURE 3 *Brämabuel avalanche area and longitudinal profile.*
Reproduced from Martinec and de Quervain (1975)

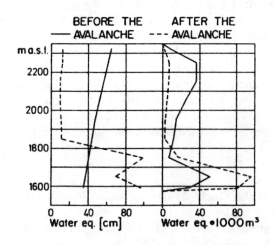

FIGURE 4 *Distribution of snow before and after the avalanche*

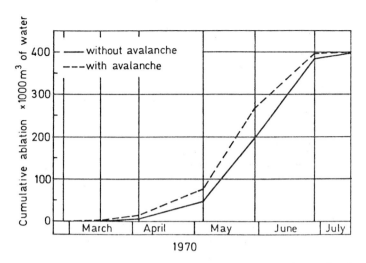

FIGURE 5 *Cumulative snow ablation without and with the avalanche.*
Reproduced from Martinec and de Quervain (1975)

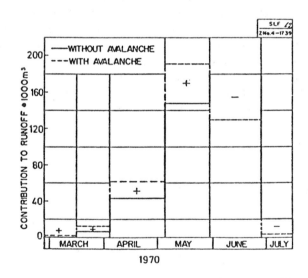

FIGURE 6 *Runoff contributions from the avalanche area with and without*
the avalanche

avalanches was about 10 times greater than the evaluated effect of the single
avalanche. It was automatically included in the measured runoff from the
Dischma basin. If it is subtracted, a fictitious runoff pattern is obtained which
would have taken place if these avalanches would not have occurred. Both
alternatives are compared in Figure 7. The redistribution of the snowmelt

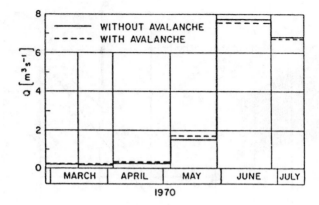

FIGURE 7 *Assumed seasonal redistribution of runoff from the Dischma basin by avalanche effects in 1970 (Martinec, 1985). Reprinted by permission of John Wiley & Sons Ltd*

runoff by avalanches reduced the discharge peak in June so that the effect can be considered as favourable, but it is rather small (Martinec, 1985).

Other studies (Iveronova, 1966; Sosedov and Seversky, 1966) report avalanches with snow deposits concentrated on small areas and, also in agreement with Figure 1, a delayed runoff as a consequence. Although this effect was opposite to that in the Dischma basin, it was considered as favourable in the given hydrological conditions because it reduced the spring floods and improved the low runoff in the summer. The authors also mention a reduction of water losses due to evaporation of snow as an accompanying phenomenon.

River Flow Detention by Avalanche Deposits

This effect is limited to avalanches which cross a water course so that the flow is temporarily dammed by snow and debris. The Brämabuel avalanche of 25 February 1970 provides a small example of this situation as shown in Figure 8. Another avalanche crossed the Dischma brook on 17 May 1979, 30 m above the hydrometric station and filled the channel with snow to a length of 70 m. As shown in Figure 9, the water stage dropped abruptly by 75 cm due to the damming which left only 8 cm of water depth in the artificial channel of the hydrometric station. Rapid oscillations of the water stage followed during the next four hours (Martinec, 1985).

A river damming of quite different dimensions is described by Morales (1966). In January 1962, the Santa river in Peru was temporarily blocked by an ice and rock avalanche which deposited altogether 13×10^6 m^3 of material,

FIGURE 8 *Damming of the Dischma brook by the Brämabuel avalanche,*
15 May 1970

of which 20%–25% was ice. When the waters of the Santa river eroded this
dam and broke through, it is reported that the water level in the river channel
rose by 10 m, caused flooding and destroyed two bridges. The other effects of
this avalanche, notably the loss of 4,000 human lives, were of course still more
disastrous.

Large water volumes may be also detained by a combined effect of an
advancing glacier, ice and snow avalanches. The advances of the Giétroz
glacier in the Western Swiss Alps caused such temporary damming of the river
Drance de Bagnes several times in the history. In June 1818 an artificial lake,
3 km long, 200 m wide and 70 m deep, was created. In spite of protective
measures, a volume of 20×10^6 m^3 of water broke through and the floodwave
travelled towards the Rhône valley. The whole area was devastated and 50
human lives lost (Bachmann, 1978). A repetition of flooding is now curbed by
the Mauvoisin dam. At the same time, the Giétroz glacier has to be
permanently observed in order to prevent a floodwave in the Mauvoisin
reservoir being caused by falling ice masses.

A long-term damming with the danger of subsequent flooding can be caused
in particular by landslides. These phenomena exceed the scope of this paper.

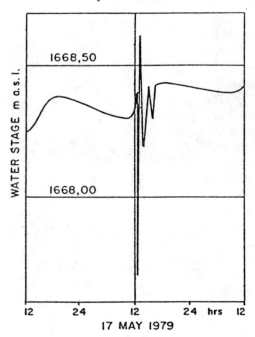

FIGURE 9 *Quick oscillations of the hydrograph caused by temporary damming of the Dischma brook by an avalanche (Martinec, 1985). Reprinted by permission of John Wiley & Sons Ltd.*

Conclusion

Due to snow avalanches, the snowmelt runoff in mountain basins is either accelerated or delayed and this effect may in both cases be favourable or unfavourable, depending on local conditions. The change is in most cases not very drastic because only a certain percentage of the basin area is affected by the dislodgement of snow.

A temporary damming of rivers by large avalanches containing solid material involves risks of catastrophic flooding, but luckily it is a far less frequent phenomenon.

REFERENCES

BACHMANN, R.C. (ED.) (1978) *Gletscher der Alpen* (Glaciers of the Alps), Hallwag AG Bern, 110–111

IVERONOVA, M.I. (1966) Le rôle hydrologique des avalanches. *Proceedings*

Symposium on Scientific Aspects of Snow and Ice Avalanches (Davos, 1965), IAHS Publication No. 69, 73–77

MARTINEC, J. (1971) Auswirkungen der Brämabuel-Lawine 1970 auf den Verlauf der Schneeschmelze (Effects of the Brämabuel avalanche 1970 on snowmelt). Internal Report No. 514, SLF Weissfluhjoch/Davos, 40 pp

MARTINEC, J. (1985) Time in hydrology. In: *Facets of Hydrology II.* Rodda, J.C. (Ed.) John Wiley & Sons, Chichester – New York – Brisbane – Toronto – Singapore, 262–263

MARTINEC, J. and DE QUERVAIN, M.R. (1975) The effect of snow displacement by avalanche on snowmelt and runoff. *Proceedings of the Moscow Symposium* 1971, IAHS Publication No. 104, 364–377

MORALES, B. (1966) The Huascaran avalanche in the Santa Valley, Peru. *Proceedings Symposium on Scientific Aspects of Snow and Ice Avalanches* (Davos, 1965). IAHS Publication No. 69, 304–315

SOSEDOV, I.S. and SEVERSKY, I.V. (1966) On hydrological role of snow avalanches in the northern slope of the Zailiysky Alatau. *Proceedings Symposium on Scientific Aspects of Snow and Ice Avalanches* (Davos, 1965). IAHS Publication No. 69, 78–85

Hazards to Water Resources Development in High Mountain Regions. The Himalayan Sources of the Indus

KENNETH HEWITT

Snow and Ice Hydrology Project,
Wilfrid Laurier University, Waterloo, Canada

Introduction

IT IS still not unusual for high mountains to be viewed primarily as remote, inhospitable lands where a few traditional societies survive or as more or less untouched wildernesses visited only by mountaineers or natural scientists. That is true of hardly more than a handful of the most barren fastnesses or those deliberately set aside. Rather, nearly all mountain areas have been and are being subjected to increased resource exploitation, and integration into national and international economies. Even the highest and least populated areas are visited and their resources developed for the benefit of surrounding lands. Among the most widespread and important are developments based upon mountain water resources.

In general, mountainous watersheds are the more humid in a given region. Throughout much of the sub-tropical dry zones and in continental interiors, they are the only areas of annual moisture surplus and more or less perennial surface streams (Figure 1; Table 1). Surface and underground flow(s) to surrounding areas provide water supplied to much larger regions than the mountains themselves. Along such rivers as the Nile, Indus, Tigris-Euphrates, Amu Darya, Orinoco or Colorado, hydrological conditions in the mountain headwaters dominate flows and sediment transport for a thousand kilometres and more beyond. The behaviour of smaller rivers is often even more directly and sharply determined by what happens in the highlands. Increasingly, this runoff downstream and water in the mountains themselves are being harnessed. Large engineering works for irrigation, power generation and

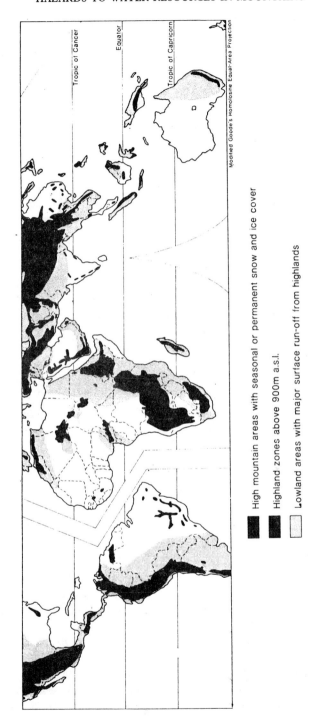

High mountain areas with seasonal or permanent snow and ice cover

Highland zones above 900m a.s.l.

Lowland areas with major surface run-off from highlands

FIGURE 1 *Worldwide distribution of high mountain zones*

TABLE 1 *High mountain water resources. A shortlist of nations or parts of nations in the tropical and sub-tropical zones, with moisture surpluses or runoff from mountainous areas. Those in italics also include significant moisture storage in and/or runoff from melting of snow and ice (Figure 1)*

THE AMERICAS	22. Somalia	39. *Pakistan*
1. *USA* (South West)	23. Tanzania	40. *India*
2. *Mexico*	24. Uganda	41. *Nepal*
3. Guatemala	25. Rwanda	42. *Bhutan*
4. El Salvador	26. Madagascar	43. *Bangladesh*
5. *Colombia*	27. South Africa	44. *China*
6. Venezuela	28. Lesotho	Xinjiang Province
7. Ecuador		Xizang Province
8. *Peru*	ASIA	(Tibet)
9. *Bolivia*	29. Turkey	
10. *Argentina*	30. Syria	SOUTHEAST ASIA
11. *Paraguay*	31. Israel	45. *Burma*
12. *Chile*	32. Yemen	46. Thailand
13. *Brazil*	33. South Yemen	47. Kampuchea
	34. Oman	48. Laos
EUROPE	35. *Iraq*	49. Vietnam
14. *Spain*	36. *Iran*	50. *Japan*
15. Portugal	37. *USSR*	51. South Korea
	(i) Azerbaijan SSR	52. *North Korea*
AFRICA	(ii) Armenian SSR	53. Philippines
16. *Morocco*	(iii) Turkmin SSR	54. Indonesia
17. Algeria	(iv) Uzbek SSR	
18. Egypt	(v) Jajik	AUSTRALASIA
19. Sudan		55. *Australia*
20. Ethiopia	SOUTH ASIA	56. *New Zealand*
21. *Kenya*	38. *Afghanistan*	57. Papua New Guinea

supply to cities and industries are used to control and redistribute the waters. Other more or less major developments are associated with tourism and recreation. Here, the attractions of mountain streams, lakes, forests, and winter snows – features that depend upon higher humidity and mountain hydrology – are exploited.

Human settlements and resource development that depend upon high mountain water supplies must also cope with a range of distinctive hydrological conditions and hazards that come with these benefits. The moisture stores and surpluses in many mountain areas tend to be fairly reliable. They often offset year-to-year variations in climatic factors. However, stream hydrographs are usually highly variable from season to season and even diurnally. This is most pronounced in those mountains whose elevation makes part or all of them subject to frost climate conditions and hence, the occurrence of snow and ice. These will be the main concern of this paper.

The occurrence of seasonal snowpacks, or perennial snow and glacier covers, can result in great seasonal variability in water yield. This is a function of the pattern of freezing and melting conditions. Where the meltwaters are a significant fraction of flow, rivers exhibit the distinctive features of nival (snow) and/or glacial regimes. These involve a marked melting season, and low flows at other times. Locally, short-term and diurnal freeze-thaw cycles also produce marked fluctuations in water yield. As a result of orographic forcing of precipitation, steep slopes, steep stream gradients and the occurrence of narrow, gorge-like sections, severe and sudden flood events are a feature of mountain rivers. Where snow and ice are involved, the intense radiation at high altitudes, alternating with the greater cloudiness of the mountains, also generates patterns of high and low flows, responding to weather conditions in the melting season. These natural aspects of high mountain hydrology have been the subject of considerable research (Glen, 1982; Young, 1985). Less often are they examined in terms of water resource development.

Mountain environments are associated with distinctive moisture regimes. They also involve a range of other processes and conditions that may interfere with water supply or damage activities and property. Broadly, these hazards derive from the influence of altitude, steep slopes and rugged topography; strong variations in climate or weather over time and space; and seismicity. They are realized in a range of more or less extreme surface processes that may give rise to destructive events. These include large landslides, avalanching and other potentially very destructive, mass movements; natural damming of rivers by landslides, vegetation, glaciers and moraines; outburst floods from the breaching of such dams; and generally higher levels of earthquake activity. As natural processes, these too have been widely studied (Voight, 1978; Price, 1981).

Reports from many mountain regions also suggest that human activities, in the way they interact with and magnify the effects of mountain conditions, are increasing environmental risks. Such problems seem to follow from abuses involving deforestation, overgrazing, accelerated erosion, and the disturbance of traditional land uses. Many are directly the result of planned, but environmentally unsound development. Damages of mining, construction, and pollution are widely observed. There are also problems due to the impact of technologies imported from other environments, and visitor activities. A considerable literature also addresses these problems (UNESCO, 1974; Ives and Messerli, 1981; Allen, 1987). Their relation to water resource development and natural disasters are less often addressed. That is the main concern here. Again the emphasis is upon region and problems where high mountain snow and ice are involved.

297

Snow and ice are, of course, almost universally considerations in the mountains of higher latitudes. Historically, most research and the more direct uses of this as a resource, have occurred there (IAHS/WMO, 1972; Roethlisberger and Lang, 1987). At progressively lower latitudes, mountains must, as a rule, reach increasingly great elevations for snow and ice to be important, or to occur at all. In detail this is modified by atmospheric circulation. The distribution of land and sea, and the topoclimatic effects of mountain barriers themselves depress or elevate the incidence of snow and ice. Consequently, tropical and sub-tropical mountains where snow and ice are present are at relatively great elevations and often extremely rugged. Perhaps, for that reason, few were directly involved in water resource development until very recently. Over the past decade or two, however, engineering works to directly harness waters in or at the foot of these high mountains have become ever more widespread. So too has the demand for understanding and monitoring high mountain hydrology for development purposes.

There are some 30 nations in the lower latitudes whose mountains have more or less substantial snow and ice (Table 1). Many are now looking to or have already begun exploiting that. The remainder of this paper will focus upon a particular case study, in which national water resource development has moved towards a massive and primary dependence upon high mountain water supplies. The larger part of them derives from melting of snow and ice.

The Upper Indus Basin and Water Development in Pakistan

Pakistan has a predominantly agrarian society with a population of about 85 million persons. Most are dependent upon the irrigated agriculture of the Indus Basin. Here is the world's largest integrated irrigation system, developed over the last 100 years or so. It is fed through 16 diversion dams (barrages) and 580 km of link canals which connect the western rivers under Pakistani control, with eastern rivers whose upstream areas are controlled by India. In addition, the system has major storage reservoirs, especially the Mangla and Tarbela dams in the Himalayan foothills.

The northern mountains of Pakistan provide the only areas of the country with substantial precipitation and an annual moisture surplus. Most of the remainder of the country has low precipitation and a water deficit in most or all months of the year. Historically, surface water supply in Pakistan depended more upon the easterly Indus streams and rainfall, both mainly deriving from the summer monsoon in Pakistan. Since the Partition of British India, in 1947, the main surface water development projects have concentrated upon the western Indus streams. This has involved power generation, irrigation, and as noted, feeding of the waters eastward through the replacement works. Meltwater from Himalayan snow and ice dominates the flow of these western rivers, especially the main stem of the Indus.

The Upper Indus Basin (UIB) encompasses an area of some 250,000 km^2 in a mountainous region of the Western Himalayan, Karakoram and Hindu Kush Ranges (Figure 1). It is a basin having a globally extreme degree of ruggedness and range of relief. At Mandori gauging station (Attock), where the river emerges from its mountain headwaters and is joined by the Kabul river, more than 75% of the flow derives from snow and icemelt. Total yield here is, on average, 98×10^9 m^3 yr^{-1}. Preliminary estimates indicate that snow and ice meltwater supplies 75% inflow of the Kabul at Warsak, 80% of Swat river, 85% of the inflow of the main Indus to Tarbela Reservoir, and 65% of the Jhelum at Mangla Reservoir. In proportion and scale, these are exceptional snow and ice regimes for rivers in the semi-arid subtropics. At the same time, problems of monitoring, modelling, forecasting and hazards parallel those in other high mountain basins, especially of the Central Asian and Andean ranges (Hewitt, 1985).

The level of water development now achieved means that almost all of the flow of these rivers, in nearly every month, is controlled and utilized. In particular, total flows from mid-September to mid-June are exploited in most years. On the other hand, despite enormous investment in reservoirs, storage is limited, and dependent upon high flows in July and August. As a result, the economy has become increasingly subject to problems of river-flow variability. Only now are the implications beginning to be recognized.

It should be noted that *peak* flows from snow and glacier melting coincide with the monsoonal period. This coincidence of heavy rains in parts of the Plains and Himalayan foothills, masks the major role of meltwater even then. Being the most common source of flood peaks, monsoonal rain storms also mask the large role of high meltwater flows upon which those peaks are superimposed. Finally, snow and ice regime rivers are less episodic than monsoon rains. Often, therefore they compensate for vagaries of the monsoon, but sometime they reinforce them.

Nevertheless, meltwater yields are variable. Variations in their contribution to flow in generally low-flow months, and the timing and rate of rise of spring flows, are particularly crucial. Given the level of use, these variations now appear in occasional severe shortfalls, notably in spring and early summer. Then, as in 1985 for instance, several months of electricity 'load shedding' and irrigation supply shortfalls become serious national concerns.

Hydroclimatic conditions in the Upper Indus Basin

The main concerns here are hazards to water development. However, these can hardly be understood without reference to the conditions governing the yield of water from the mountainous headwaters of the Indus.

There have been gauging stations on the major tributaries of the Upper Indus, and its main stem for several decades, at least. The hydrographs from such stations leave no doubt as to the extreme seasonal variability of flow

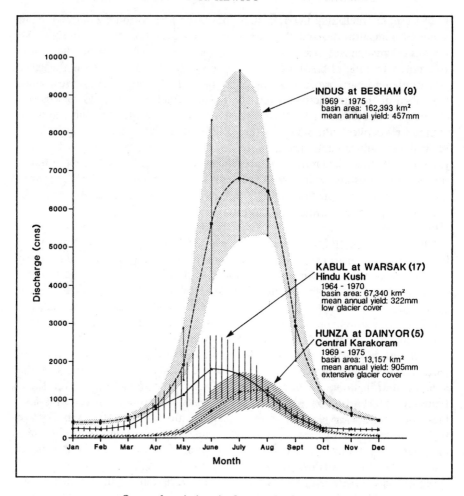

FIGURE 2 *Seasonal variations in flow on the Indus and its tributaries
(drafted by G. Young)*

(Figure 2). Between two-thirds and nine-tenths of the water yield occurs in just six to 10 weeks of the year. Depending upon the elevation and location of the catchments, some of these rivers are largely snowmelt waters, some largely glacier melt, and others a mix. That is to say, they come within the 'Highland Snow' and/or 'Highland Ice' regimes. Some of the southerly basins of the Himalayan Front Ranges are what Pardé termed 'nivo-pluvial' or 'transitional nival', having a higher fraction of monsoonal and winter rains contributing to runoff (Beckinsale, 1969).

Of decisive importance is the nature and influence of an exceptionally high and rugged mountain environment. Topography largely determines where snowfall occurs, the relative accumulations of snow and ice, and their patterns of release by melting. Two outstanding factors are the great variation of snowfall with altitude, and the altitudinal migration of melting temperatures over the hydrological year (Figure 3). These two factors combine to ensure that only a fraction of the whole Upper Indus Basin – probably less than 30% – contributes perhaps more than 80% of the river's flow. In essence, all the hydrological 'action' takes place in the zones above 2,500 m, and for most of the basin over 3,000 m asl. Most of the actual yield of water to streams derives from a belt between altitudes of 2,500 and 5,500 m. Above that again, there is often heavy snowfall and large stores of snow and ice, but little melting. Below 2,500 m there is generally little precipitation and high evaporation in the interior valleys, where rain shadow effects and desiccating valley wind systems create an arid sub-climate (Hewitt, in press).

At the end of the winter season, the mountain regions of the UIB have an extensive seasonal snowcover. An area of the order 200,000 km^2 is usually involved. For the Indus main stem, snow may cover more than 90% of the catchment above Tarbela Dam, and commonly more than 70%. In a poor year, it may, however, be less than 60%.

From May through July, melting of this seasonal snowcover provides the bulk of the flow of the Indus streams. By late July, when most of the region is yielding peak flows, on average the snowcover has been reduced to a small fraction of its winter extent (10–30%). Moreover, the bulk of the actual snow cover disappears before significant rises of the rivers are noticed. This reflects the extreme gradient of precipitation with elevation. The deep snow packs that yield most of the meltwater only comprise a fraction of this; essentially in areas between 2,500 and 5,000 m asl. Melting of the lower, thinner snow cover may be critical for spring sowing and when reservoirs are low. The main melt is usually in progress by late April. Snowmelt dominates flow from the whole basin until early July.

The geography and area-altitude relations of snowfall and snowmelt are important considerations. Although seasonal snowmelt is a contributor to all northern rivers, it is dominant for those draining the main Himalayan ranges. These are the ones lying roughly south of the axis along the Gilgit river, and the main Indus from where it is joined by the Gilgit to Skardu. North of this, in the Karakoram head waters, glacier basins are the larger contributors.

Estimates of the area of perennial snow and ice cover are between 20,000 and 25,000 km^2 or nearly 10%. There are thousands of glaciers, but the 35 or so largest ones – some like the Siachen (1,200 km^2) and Biafo (640 km^2) being of very great size — dominate the hydrology. The glaciers originate in high snowfall areas, generally above 4,800 m asl. Most of the water yield is from ice that has flowed to lower altitudes, mainly in the range 3,000–5,000 m asl, and during an intense period of melting that usually dominates river flow from mid-July until early September (Hewitt et al, 1989; Ferguson, 1984).

Geographically, the most important area is the immense arc of permanent snow and ice stretching from Tirich Mir (7,690 m) in the Hindu Kush, through K2 (8,611 m) in the Karakoram, to the Saser Muztagh in the

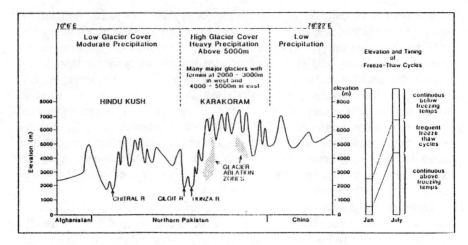

FIGURE 3 *West-East transect at 36° N latitude*

southeast. This is roughly 900 km, or 6 degrees of longitude. From late summer satellite imagery, one sees that this is the largest concentration of perennial snow and ice on the continent of Asia. The area is about 15,000 km². It constitutes an immense store of fresh water. However, more significant to our present purposes is the rate of supply, throughput and melting of this snow and ice, and the relative role of seasonal snowcover that develops on and around it.

The bulk of the snowfall nourishing the glaciers derives from westerly sources during the winter (rabi) half of the year. However, even in summer the predominant mode of precipitation is also snowfall. This may be a large fraction of the total supply. Studies indicate that it has provided about one-third of the snowfall nourishing glaciers in the last decade. This also seems to come predominantly from the west rather than monsoon (Wake, 1987). Altitudinal differences in precipitation are much more marked in the northern than in the southern part of the UIB. Rain shadowing by the surrounding high ranges, and the powerful valley wind systems result in severe desiccation of the valleys. This is most marked below 3,000 m asl in the westerly, and 4,500 m in the easterly parts of the Karakoram. There are also marked differences in the degree of aridity and its upper limits on south-facing as against north-facing slopes. The latter often have snowlines as much as 500 m lower than adjacent south-facing slopes.

The valleys around Gilgit and Hunza, for example, are among the driest areas of Central Asia. The upslope moisture gradients involve changes from less than 100 mm annual precipitation on the valley floors, to more than 1,500 mm over an altitudinal range of 4,000–5,000 m and horizontal distances of 10–20 km or less. Thus, the precipitation-enhancing and shadowing effects of the main mountain ranges provide dramatic contrasts that greatly complicate the hydrological picture.

Mountain environments generally are distinguished by more or less strong variations with altitude, aspect and, in extensive mountainous areas, from one mountain range or valley, to the next. This is usually apparent in precipitation patterns, wind and avalanche activity, seasonal and perennial snowcover, degree of glacierization, and patterns of runoff. Such variability is typical of the UIB and to an exceptional degree.

The meltwater that supplies the larger part of the Indus' flow consists, as we have seen in part of area snowmelt, including extensive basins with little or no perennial ice, and the ablation of glacier ice. While there are certainly important variations between areas of snowmelt and glaciers in different parts of the UIB, it is of first importance to differentiate between the snowmelt and glacier ablation components. These involve differing conditions and patterns of melting. They may respond quite differently to the seasonal climatic conditions in the basin. Individually, snowmelt or icemelt in mountain regions show large year-to-year variations, not uncommonly as much as 30% in various parts of the world. They are also commonly out of phase with each other, such that their relative contribution to runoff also varies greatly from year to year. In fact, a frequent situation is for glacier ablation to compensate for poor snowmelt and vice versa. As a result, runoff from snow and ice regime basins tends to fluctuate *less* than the precipitation conditions alone would indicate. However, it is not always like that. The more serious problem of water resources planning is that sometimes both poor snowmelt and poor glacier ablation, or high yields from both coincide. These are the 'drought' and 'flood' conditions respectively of snow and ice regime rivers.

Recent studies elsewhere indicate that mountain glaciers contain larger stores and are responsible for a larger share of freshwater supply than had been thought, before exact and comparative work in the International Hydrological Decade. In the case of the UIB their significance is out of proportion to their areal extent for at least two reasons. Firstly, glacier basins coincide largely with the areas of high precipitation, whereas most of the non-glacierized area is semi-arid or subhumid. Secondly, the Indus Waters Treaty, as noted, requires Pakistan to rely increasingly upon flow from the more northern and western parts of the UIB. Here monsoonal rain is the minor contributor and more of the non-glacierized area is arid.

The net effect of these topoclimatic conditions and their variable expression in the distribution of snow, ice and meltwater yields, is to produce a complex, often highly fluctuating water resource. A singular problem and source of risk for water development is inadequate records, monitoring and research into the conditions controlling runoff here. Until now, there have been few efforts to determine and distinguish the roles of seasonal snowmelt as against glacier melt, although these are significantly different in their roles. There had been no means to provide managers with data from the critical attitudinal zones between 2,500 and 6,500 m above sea level from whence nearly all the snow and ice melt derives. There have been some satellite-based studies of the UIB forecasting. These are important but inadequate without better knowledge and ground control.

Here too are some common problems for water development in high mountains. A more or less variable and sharply fluctuating water supply, comes from regions often remote from and unfamiliar to agencies responsible for water management. The mountain environment itself has rarely received much attention, and lacks those records of climate and streamflow taken for granted in many other areas.

Hydrological risks

The most important concerns of most water resource developments relate to water supply and the risks of shortfalls in supply. There are shortfall risks associated with the moisture stream in most environments. They are all the more severe when depending upon mountain rivers, especially without adequate monitoring, and knowledge to forecast their behaviour.

However, in high mountain environments such as those of the UIB, a range of other potentially destructive conditions interfere with the moisture stream. They threaten human settlements and installations. Many involve snow and glaciers. They derive in part from the same conditions as govern water supply itself. Some will directly interfere with or damage a monitoring system and installations for water and power development. Some may spread their devastation to the plains, or indirectly cause problems there. To the extent that these hazards do involve snow and ice, they will be described here.

(1) *Avalanches.* A large fraction of all the snow that falls in the UIB is avalanched through some hundreds of metres. This also relates to the altitudinal relations of the snowfall. Heavier falls occur at higher altitudes where slopes tend to be steeper and snowpacks more unstable. Most Karakoram glaciers are predominantly nourished by avalanches. Where settlements, roads and other installations occur, avalanche damage to life and property is a serious problem. Roads in Kaghan, Swat, Gilgit, Hunza and Baltistan at Azad Kashmur are often blocked or closed for much of winter, because of avalanching (deScally and Gardner, 1986).

(2) *Glacier dams.* Glacier dams and outburst floods ('jokulhlaups') have been reported in many glacierized mountain regions, and may create hazards for human populations. Unusually large and dangerous cases are those where the ice blocks rivers draining large areas. These have been exceptionally common in the Karakoram. Thirty-seven destructive outburst floods have been identified for the Karakoram region in the past 200 years (Hewitt, 1964, 1982). Thirty glaciers are known to have advanced across major headwater streams of the Indus and Yarkand rivers. There is unambiguous evidence of large reservoirs impounded by 18 of these glaciers.

Meanwhile, a further 37 glaciers interfere with the flow of trunk streams in a potentially dangerous way.

Geographically, glacier dams in main river valleys have occurred from the far western to the far eastern part of the Karakoram range. They are recorded in the Lesser Hindu Kush, Nanga Parbat, Haramosh, Hindu Raj, Aghil, and far northeast Hindu Kush ranges. These are areas where maximal local relief for the Earth's land surface is approached. Altitudinal differences of some 3,000–5,000 m over distances of 10–30 km from main river valleys, are reflected in the steep fall of many of the glaciers that cause problems. On the other hand, the trunk streams of the rivers are deeply incised and fall relatively gently. Steep, glacierized valley walls and tributary valleys may thus cut off extensive drainage areas. High relief, steep average fall and large climatic gradient between upper and lower reaches promote vigorously active ice masses in the region. This is reflected in those few for which movement data had been collected and which flow 150 to 250 m a year. All except two of the glaciers known to have formed dams flow in northerly and easterly directions. This suggests that aspect is a major factor for the potential damming, as it is in other characteristics of these high, sub-tropical glaciers. While dams of at least four glaciers have repeatedly reformed over periods as long as two decades, the lakes rarely last more than two summers without an episode of complete draining. It is the sudden draining of such lakes that represents the greatest hydrological hazard. However, the ponding and ice barrier itself may also interfere with routeways or destroy grazing land. None of the dams has occurred in the zone of permanent habitation. Before turning to the outburst flood problem, another source of damming may be noted.

(3) *Large landslides.* Although not directly a part of the moisture stream, mass movements on slopes respond to moisture supply and may interfere with water flow or access to water resource installations. The steep slopes of the Karakoram are mostly bare of vegetation. This is due to aridity or centuries of grazing by domestic animals at altitudes below 4,000 m or so, and to cold, snow and ice above that. As a result, routeways and those settlements and installations not sited with care, are subject to a range of landslide problems. At lower elevations, one of the most common and severe hazards derives from mudflows and debris flows. Below 4,000 m asl are commonly enormous, unconsolidated deposits resulting from Pleistocene and recent processes. In spring and early summer, these poorly vegetated materials may be saturated with snowmelt water. Vigorous episodes of melting, especially of avalanched snow in gullies, produces many large mudflows. Rare, concentrated rainstorms in summer may have the same effect (Miller, 1984, vol. II, chs 17–21). Occasionally, these events are large enough to dam the rivers. This happened to the Gilgit

river in Gupis, July 1980 (Hughes and Nash, 1986). A mudflow of great size and power filled the valley with a lobe some 500 m wide, and deep enough to back the water in a lake some 6 km long. A village and most of its fields were drowned, and the highway blocked.

Large catastrophic landslides also occur, especially on the slopes of the main glacially oversteepened valleys and of the major river gorges. Along most of the rivers are high terraces in old moraines and fluvial deposits, and alluvial fans from side-valleys. The cliffs in these poorly consolidated materials are commonly more than 100 m high. Suites of terraces along the main Indus stem and large tributaries below about 2500 m as extend over 1000 m in elevation. Not surprisingly, slumps and bouldery earth falls are recurrent problems here, blocking highways and other access routes that must pass along gentler valley floor areas beneath.

At higher elevations, large rockslides present a severe hazard. In July 1986, three catastrophic rockslides descended onto one of the glaciers that are being monitored in the Karakoram. They involved 20 million m^3 of rock, and covered 4.1 km^2 of the glacier with a layer 6 m thick (Hewitt, 1988). One of the stake arrays for movement and ablation studies was buried. Later, deposits from similar events in the past were discovered in the basin, suggesting these are recurring events. This particular example caused a sudden 'surge' and superficial break-up of the glacier, making travel over it between villages and mountain pasture impossible for several months. The surge caused ponding of temporary ice marginal lakes and a series of minor flood waves.

The worst episodes of landsliding are associated with earthquakes. Many hundreds or thousands of landslides have occurred in recent earthquake disasters, notably that of December 1974, around Pattan in the Indus Gorge (Hewitt, 1976). These are a threat to local access and power corridors through the mountains. In 1841, a massive landslide was triggered by an earthquake in the Indus Gorge north of Besham. The debris dammed the river for six months, creating a lake over 60 km long. This lake drained in 24 hours, creating the greatest known flood on the Indus. These hazards are relatively rare but, when they occur, present risks to population. Even the largest, best-designed hydraulic installations would demand great efforts to prevent a catastrophic outburst.

(4) *Outburst floods.* Breaching of a natural dam creates a sudden, short-term increase in discharge downstream. If breaching is catastrophic, as so often in the Karakoram region, the impact of the flood wave can far outweigh that of other high flows. This can involve a concentration of flow in the upper reaches of the rivers well in excess of weather-produced extremes of runoff, including those from snow and ice melting.

The significance of floods generally lies in the risk to human communities or installations, and also in their role as agents of erosion and sedimentation. In the case of damburst floods, over much of their course in the mountains, records indicate that they reach heights well above peak discharges from summer melting. Their dynamic character greatly magnifies their erosional competence and capacity. The 1929 outburst of Chong Khumdan Glacier deepened the Shyok river at Khalsar by more than 3 m, and the Indus at Attock by at least as much. This outburst flood was monitored from near the glacier and for more than 1,500 m downstream before it ceased to be dangerous. Gunn (1930) estimated the reservoir to have contained almost 13.5×10^8 m^3 (1.1 million acre-ft). Some 3×10^5 m^3 of ice were also carried with the flood and stranded on large blocks on the valley floor below the dam. If loss to channel storage and seepage was somewhat greater than gains from inflows below the dam, the complete draining in 48 hours suggests an average discharge between Sasir Brangsa and Khalsar in the region of 7,100 m^3 s^{-1} (250,000 cfs). However, in the steeply rising and falling main flood peak, water discharges in excess of 22,650 m^3 s^{-1} (800,000 cfs) are indicated. That equals the largest discharges measured for the entire Upper Indus at Attock. The Upper Shyok drains less than 2% of the basin, and most of its area is arid (Hewitt, 1982). There is evidence of some 35 disastrous jokulhlaups since 1826. The rarer landslide dams have resulted in the largest dam-burst floods.

(5) *Surging glaciers.* Some ice dams may have been the result of glacier surges. These events can produce an advance of a glacier of several kilometres in a few months. A surge is commonly accompanied by increased water and sediment discharge. It can be extremely hazardous to settlements or installations in its path. At least 11 surges of exceptional scale have been recorded for the UIB (Hewitt, 1969). In 1953, the Kutiah Glacier in the Haramosh Range advanced 12 km in two months. Indications are that many more than these 11 glaciers in the region may surge. Some, like the Hassanbad in Hunza, are a serious threat to human activity.

As noted, one of the glaciers monitored, the Bualtar in the Central Karakoram, 'surged' in 1987, accelerating from a flow rate of about 0.6 m day^{-1}, to over 7.0 m day^{-1}. This was directly related to the large rockslide. It caused very severe problems of erosion, and destruction of routeways for villagers in Nagyr.

(6) *The range of risks.* Figure 4 shows the distribution of known hazardous conditions or events that have affected the Upper Indus drainage. It shows clearly the widespread and varied incidence of problems directly relating to the high mountain environment. However, it seems fair to say that the human risks from most of these events were, until recently, confined to impacts upon the local, high mountain communities. Only the very latest floods seriously affected the lowland areas and major centres of population. With increased development and integration of the northern

307

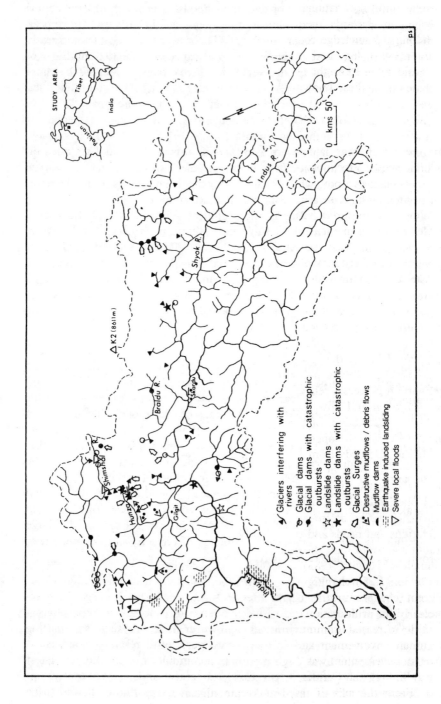

FIGURE 4   Distribution of known hazards that affect Upper Indus drainage

areas into the national economy, the scope of the risks from these mountain hazards increases and changes. This raises great need for technical knowledge concerning the hydrological processes, much greater information on and monitoring of the high mountain areas and considerable foresight in planning water developments so as to minimise losses from these known hazards.

Concluding Remarks

In the image that late twentieth century nations like to project, it is economic growth, technology and industry that receive the greatest attention. Yet, the demands upon, and exploitation of, planetary resources has never been so great. The risks associated with these demands and the environments providing the resources, are growing rather than declining problems for science and government. Larger dependency on a particular resource creates larger risks, or at least, puts more people at risk. Interruptions or shortfalls affect people and economies more or less far removed from the source area. In the case of mountain waters, they include the often more densely populated surrounding areas. Meanwhile, a range of new pressures and hazards associated with exploitation develop in the mountain lands.

It is important to recognize that these are hazards of development, rather than intrinsic features of the high mountain environment. Too often, the impression is given that problems of flood, drought, earthquake or landslide are simply the result of natural extremes and 'unstable' environments. To be sure, natural processes reflect the controlling conditions in a given habitat. But whether, when and how they become risks or damage to human activity is essentially a function of that activity. Avalanche risks or costs change dramatically with the opening up of a high valley to all-year-round traffic and ski-tourism. Hazards from snow- and ice-fed streams are completely remade when dams are built or newly irrigated lands made dependent on them.

We have noted a broad range of potentially hazardous processes associated with Upper Indus Basin hydrology. There is a history of past damages caused by them. However, an integral factor of risk is the circumstances the changing expectations and needs and the 'stability' of the human communities involved. Without a grasp of that, the study of objective hazards is of limited value if not meaningless (Hewitt, 1983). It becomes the more so, in societies subject to more or less rapid change.

Recent changes in the Upper Indus Basin involve some large planned projects and a proliferation of small-scale aid and intervention. There is also generally increased communications and numbers of visitors, expanding government involvement and access to modern technology. The single most important development was the completion, a decade ago, of an all-year, paved highway linking the Central Karakoram valleys to the plains of Pakistan and China. From the axis of the Karakoram highway, especially, pressures for change have spread to the communities and economy of the surrounding valleys.

In addition to the two great dams, Tarbela and Mangla, where the Indus and Jhelum leave the mountains, there are now plans for one or more other large reservoirs in the mountains with access along the Karakoram Highway. In many of the smaller valleys, local communities and government agencies have built or are planning to build hydro-electric projects and new water supply systems. These developments are associated with changing risks from natural forces, sometimes increasing the scale of loss or redistributing it.

Some problems are a direct consequence of development works. The building of roads in this environment brings an enormous increase in the occurrence of, and effort required to deal with, landslides and related processes that block roads. Smaller hydraulic installations are often damaged or rendered useless by landslides, floods or massive build-up of sediment. But, perhaps the root problem is the lack of knowledge, experience and concern for the environmental implications evident in many of those new developments. They are built with the *benefits*, not the *risks* in mind. Conversely, mountain residents complain that the officials coming from the cities of the plains are too easily frightened away from approving developments by conditions that appear dangerous but are not. These are all elements of water resource development that include and go beyond hydroscience itself. Evidently, they are important responsibilities for science and government in regions where rapid development is taking place, and considerable risks are involved with it.

Acknowledgements

Parts of the research reported were funded by the International Development Research Centre, Ottowa, Canada; Pakistan's Water and Power Development Authority, and Wilfrid Laurier University's Office of Research.

REFERENCES

ALLEN, N.J.R. (1987) Impact of Afghan refugees on the vegetation resources of Pakistan's Hindu Kush – Himalaya. *Mountain Res Devel* 7: 200–204

BECKINSALE, R.P. (1969) River regimes. In: *Water, earth and man* Chorley R.J. (Ed.) Methuen and Co. Ltd., London, pp. 455–471

DE SCALLY, F.A. and GARDNER J.S. (1986) *Avalanche hazards in Kaghan Valley, Pakistan*. Snow and Ice Hydrology Project, Working Paper #2, Wilfrid Laurier University, Waterloo, Canada, 26p

FERGUSON, R.I. (1985) Runoff from glacierised mountains: a model for annual variation and its forecasting. *Water Resources Res.* 21 702–708

GLEN, J.W. (1928) *Hydrological aspects of alpine and high mountain areas*. International Association of Hydrological Sciences Publication No. 138, p. 350

GUNN, J.P. (1930) Report of the Khamdan Dam and Shyok flood of 1929. Government of Punjab Publication, Lahore

HEWITT, K. (1964) The great ice dam. *Indus* 5: 18–30

HEWITT, K. (1968) Records of natural damming and related events in the Upper Indus Basin. *Indus* 10

HEWITT, K. (1969) Glacier surges of the Karakoram Himalaya (Central Asia). *Canad J Earth Sci* 6: 1009–1018

HEWITT, K. (1976) Earthquake hazards in the mountains. *Nat Hist* 85: 30–37

HEWITT, K. (1982) *Natural dams and outburst flods of the Karakoram Himalaya.* International Association of Hydrological Sciences Publication No. 138, pp. 259–269

HEWITT, K. (1983) Seismic risk and mountain environments: aspects of the role of surface conditions in seismic risk. *Mountain Res Devel* 3: 27–44

HEWITT, K. (1984) Ecotonal settlement and natural hazards in mountain regions: the case of earthquake risks. *Mountain Res Devel* 4: 31–37

HEWITT, K. (1985) *Snow and Ice Hydrology in remote, high mountain basins: the Himalayan sources of the River Indus.* Snow and Ice Hydrology Project, Working Paper #1, Wilfrid Laurier University, Waterloo, Canada, 29p

HEWITT, K. (1988) Catastrophic landslide deposits in the Karakoram Himalaya. *Science* 242: 64–67

HEWITT, K. (in press) The altitudinal organisation of Karakoram geomorphology. *Zeitschrift f. Geomorphologie.* Special Issue

HEWITT, K., WAKE, C.P., YOUNG, G.J., and DAVID, C. (1989) Hydrological investigations at Biafo Glacier, Karakoram Himalaya: and important source of water for the Indus River, *Annals of Glaciology*, 3 (in press)

HUGHES, R.E. and NASH, D.F.T. (1986) The Gupis debris flow and natural dam, July 1980. *Disasters* 10: 8–14

INTERNATIONAL ASSOCIATION OF HYDROLOGICAL SCIENCES (IAHS). WORLD METEOROLOGICAL ORGANIZATION (WMO) (1972) *Role of snow and ice in hydrology.* (2 vols) Proceedings of the Banff Symposium

IVES, J.D. and MESSERLI, B. (1981) Mountain hazards mapping in Nepal: introduction to an applied mountain research project. *Mountain Res Devel* 1: 223–230

MILLER, K. (ED.) (1984) *The International Karakoram Project*, (vols I and II). Cambridge University Press, Cambridge

PRICE, L.W. (1981) *Mountains and men.* University of California Press, p. 506

ROESTHLISBERGER, H. and LANG, H. (1987) Glacier hydrology. In: *Glaciofluvial Sediment Transfer* Gurnell, A.M. and Clark, M.J. (eds). John Wiley and Sons, Chichester

SHIPTON, E.E. (1939) Karakoram. *Geographical J* 95: 49–427

UNESCO (1974) Impact of human activities on mountain and tundra ecosystems. Programme on Man and the Biosphere. Working Group Project 6.

VOIGHT, B. (ED.) (1978) *Rockslides and avalanches.* Elsevier, Amsterdam

WAKE, C.P. (1987) Snow accumulation studies in the Central Karakoram. *Proceedings of the Eastern Snow Conference (North America).*, 44th Annual Meeting, Fredricton, pp. 19–33

WORLD METEOROLOGICAL ORGANISATION (1977) *Flood Forecasting and Warning System for the Indus Basin, Pakistan.* United Nations Development Programme, Geneva, pp. 24

YOUNG, G.J. (ED.) (1985) *Techniques for prediction of runoff from glacierized areas.* International Association of Hydrological Sciences Publication No. 149, p. 148

Conclusions from the Technical Conference on Hydrology of Hazards
===

Ö. STAROSOLSZKY

President of the Commission for Hydrology,
Research Centre for Water Resources Development,
VITUKI, Budapest, Hungary

NATURAL hazards present dangers to humans and to their property. They present risks which can be high especially if they are ignored or proper precautions are not taken. However, steps can be taken to mitigate disasters, but of course even the most ambitious scheme cannot prevent or alleviate the effects of the most severe events – volcanic eruption, flood, earthquake, typhoon, drought or landslide. Water enters into many of these occurrences since too much water brings about landslides and avalanches and too little causes drought.

Though humans influence nature more and more in the present world, nature is still able to spring its surprises on us through these hazards. Society can be endangered or it can be severely disrupted and the very fabric of a region or a nation can be completely wrecked such as recently in Bangladesh. As far as water is concerned, two types of hazard can be identified that are particularly invidious: the breach of the dam wall or a reservoir and the accidental pollution of a river, lake or aquifer. There are more than 16,000 large dams on Earth; several thousands of tons of poisonous material are stored or transported continuously.

While regular natural events observed for several decades or even hundreds of years can be treated statistically, the man-made accidents are much more difficult to characterize.

The most serious cases are the coincidences of two or more hazards, and their risks can only be roughly estimated. Multilateral interactions may superpose certain phenomena which can in such a way, reinforce or reduce the components, e.g. the effect of pollution may be reduced when a simultaneous flood can dilute the pollutant extensively, or, alternatively, a rapid reduction of the flow may cause more serious consequences during the spillage of the pollutant. Traditional probability theories can hardly be applied for such conditions.

While people living in a hazard-prone area can easily be prepared for certain regular events, such a coincidence of different natural and/or man-made hazards may be catastrophic, since such rare phenomena make severe impact through their unexpectedness.

Here, however, we are considering the hydrological aspects of water-related hazards, and neglecting from the practical point of view the coincidence or joint occurrence of two or more natural and/or man-made hazards. We know, however, something about the coincidence of these events: for example, serious dambreaches often occur during floods when spillways are overtopped or high water levels occur in the reservoir and may cause serious underseepage resulting in hydraulic failure.

Landslides often coincide with extreme rainfalls, while earthquakes may be simultaneous with volcanic eruptions. Landslides may close a valley and form a reservoir. If flooding is simultaneous, the newly-formed reservoir is filled within hours or days, thereafter endangering the areas downstream.

Failures of man-made structures like roads, railroads and bridges, may have serious consequences and may upset any preparedness measures (e.g. the mobilization and transportation of forces, civil defence, the evacuation of the population and the minimizing of property risk).

While nations spend billions of dollars to improve or maintain preparedness against each other in the form of armaments, relatively little has been done in the improvement of the preparedness against natural and man-made hazards. On the contrary, by developing new devastating weapons the danger of accidental outbreak of war and nuclear pollution is increased. So, it is time to reconsider how humankind can protect itself against natural and man-made hazards, e.g.:

- by investigating the most common natural and man-made hazards, particularly the possibilities of their joint occurrence;
- by increasing the ability and capability to predict the hazard (proven) areas and to map the areas at risk (especially areas where timely prediction or warning is impossible) and to forecast the hazards in due time;
- to effectively transfer scientific and technical information about hydrologic hazards to local governmental and civil defence agencies and to the populations at risk;
- to improve co-operation and co-ordination between the scientific community and the appropriate local government and civil defence agencies for development of effective, workable response plans to hydrologic disasters. Follow-up and rehearsal of plans must continue on a regular basis;
- to improve preparedness of the population at risk and establish proper legal, financial, technical and organizational measures to reduce the harmful effects of the natural hazards and the risk of any man-made hazards;

- to develop methods, structures and equipment to be deployed to protect the endangered population and its property;
- to help developing nations, where experience and financial means are limited, to build up their preparedness and protection measures in order to prevent or, at least, localize hazards.

What conclusions can we derive from the technical conference?

Water hazards and the hydrological implications of these hazards may endanger a large amount of the world's population, especially as a consequence of rapid socio-economic development such as urbanization and industrialization. A greater proportion of the population is now endangered by natural and man-made hazards.

Recognizing and appreciating the efforts of the international organizations, like WMO, particularly in the field of floods, one may conclude that relatively little progress has been made with respect to other hazards causing hydrological problems; in particular, the study of the point occurrence of the different hazards.

Considering the rapid reply of the international organizations to the challenge of the Chernobyl accident, it seems to be important not to wait for similar serious events to stimulate the necessity of the international co-operation in other fields.

It is therefore suggested that:

- the report of the VIIIth session of the Commission for Hydrology should reflect properly the needs for action, at least within the WMO programmes;
- WMO Executive Council should taken proper measures to incorporate in WMO programmes necessary activities, in order to improve preparedness against natural and man-made hazards;
- since the UN Decade on Natural Disaster Reduction will be launched before the next WMO Congress, the Executive Council should decide how to reflect these needs in the recent programme. XI Congress may allocate necessary resources for the 1992–2000 period, and give emphasis for actions within the Third Long-term Plan.

The Technical Conference, within the limited time available, has made a considerable contribution to the development of awareness and has supplied good examples of the importance of the hydrological aspects of hazard mitigation. The papers presented at the conference introduced interesting case studies and improvements in methodology in this field. So, when expressing sincere thanks to the authors, and to the WMO Commission for Hydrology for organizing the conference, I may also express my sincere hope that continuation will be ensured and this Conference is only an introduction to certain effective measures.

The UN International Decade for Natural Disaster Reduction should properly reflect water-connected disasters, particularly floods, droughts, dam

of landslides and volcanic eruptions in order to prevent and mitigate disasters in freshwater bodies, including the failures of man-made structures due to natural disasters. Close collaboration between the water-oriented international organizations in this field may help the cost-effective utilization of the available resources. The experiences of the Technical Conference offer a sound basis to plan water-oriented activities during the Natural Disaster Reduction Decade.

Author Index

Subject Index